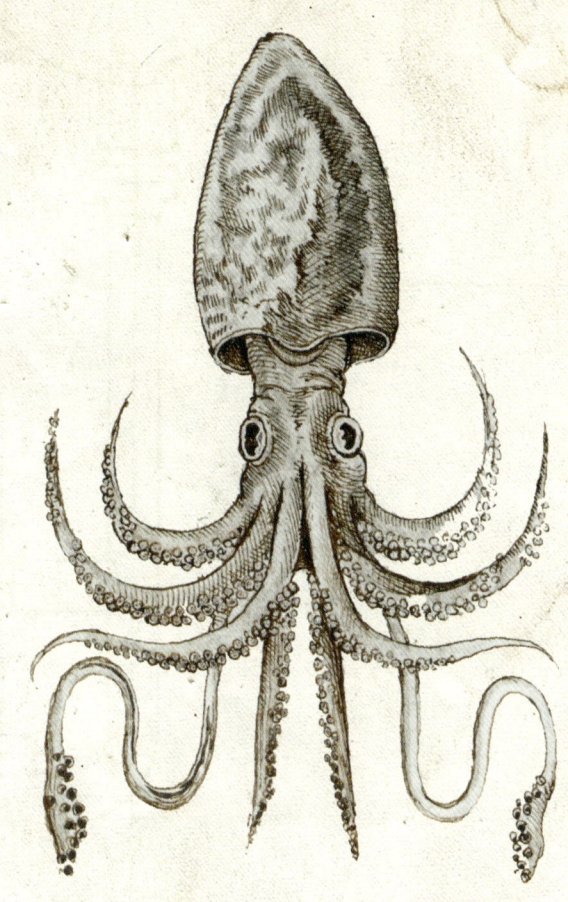

THE BOOK OF
SEA
MONSTERS

ADLARD COLES
Bloomsbury Publishing Plc
50 Bedford Square, London, WC1B 3DP, UK
Bloomsbury Publishing Ireland Limited
29 Earlsfort Terrace, Dublin 2, Ireland

BLOOMSBURY, ADLARD COLES and the Adlard Coles logo are trademarks of
Bloomsbury Publishing Plc

First published in 2025

Copyright © Prema Arasu, 2025

Prema Arasu has asserted their right under the Copyright,
Designs and Patents Act, 1988, to be identified as Author of this work

For legal purposes the Acknowledgements on page 220
constitute an extension of this copyright page

All rights reserved. No part of this publication may be: i) reproduced or transmitted in any form, electronic or mechanical, including photocopying, recording or by means of any information storage or retrieval system without prior permission in writing from the publishers; or ii) used or reproduced in any way for the training, development or operation of artificial intelligence (AI) technologies, including generative AI technologies. The rights holders expressly reserve this publication from the text and data mining exception as per Article 4(3) of the Digital Single Market Directive (EU) 2019/790

Bloomsbury Publishing Plc does not have any control over, or responsibility for, any third-party websites referred to or in this book. All internet addresses given in this book were correct at the time of going to press. The author and publisher regret any inconvenience caused if addresses have changed or sites have ceased to exist, but can accept no responsibility for any such changes

A catalogue record for this book is available from the British Library
Library of Congress Cataloguing-in-Publication data has been applied for

ISBN: HB: 978-1-3994-1452-4; ePub: 978-1-3994-1453-1; ePDF: 978-1-3994-1451-7

2 4 6 8 10 9 7 5 3 1

Typeset in Source Serif Variable by Phil Beresford
Printed and bound in Printed in UAE by Oriental Press

To find out more about our authors and books
visit www.bloomsbury.com and sign up for our newsletters

THE BOOK OF SEA MONSTERS

Leviathans of Literature

Prema Arasu

ADLARD COLES

LONDON · OXFORD · NEW YORK · NEW DELHI · SYDNEY

Contents

Introduction 6

1. Enūma Eliš (Unknown) 11
2. Odyssey (Homer) 19
3. The Aeneid (Vergil) 29
4. Metamorphoses (Ovid) 33
5. Naturalis Historia (Pliny the Elder) 39
6. The Book of Jonah (Unknown) 47
7. Beowulf (Unknown) 51
8. The Natural History of Norway (Erik Pontoppidan) ... 57
9. 'The Sea-Nymph' (Ann Radcliffe) 63
10. The Rime of the Ancient Mariner (Samuel Taylor Coleridge) ... 67
11. 'The Kraken', 'The Merman' and 'The Mermaid' (Alfred Tennyson) ... 81
12. 'The City in the Sea' (Edgar Allan Poe) 87
13. Moby-Dick (Herman Melville) 91
14. 'Leviathan' (Celia Thaxter) 97
15. 'Caliban upon Setebos' (Robert Browning) 101
16. Toilers of the Sea (Victor Hugo) 109
17. Twenty Thousand Leagues Under the Sea (Jules Verne) ... 123
18. 'In The Abyss' (HG Wells) 133
19. 'The Terror of the Sea Caves' (Sir Charles GD Roberts) ... 147
20. 'The Seafarer' (Ezra Pound) 163
21. 'The Thing in the Weeds' (William Hope Hodgson) ... 167
22. 'Sea-Heroes', 'Thetis' and 'At Ithaca' (H.D.) 181
23. 'Dagon' and 'The Temple' (HP Lovecraft) 187
24. The Maracot Deep (Arthur Conan Doyle) 205

Credits and Acknowledgements 220
Index 221

Introduction

Monsters are human creations. They are the product of our imaginations, recurring and reforming across time and space. The word monster comes from the Latin *monstrare*, 'to show' – monsters show what is within. They represent our greatest fears: fear of the unknown, fear of the dark, fear of the natural world – even fear of the self. These anxieties are symbolically expressed in the monster's fearsome appearance and its transgressive behaviours: its excess size, its claws and tentacles, its bloodlust, its violation of territories. As symbolic manifestations of abnormality, monsters are ultimately reflections of our innermost anxieties and, sometimes, desires. It is not surprising, then, that sea monsters are often present in the mythology of seafaring and coastal cultures. Great serpents and giant lobsters infest the waters of medieval and renaissance maps as a warning of what might lie in the unknown.

Stories and art about sea monsters are projections through which anxieties about the natural world can be expressed, examined and ultimately mastered. The sea monsters of stories transmitted from generation to generation are analogues for dangerous weather events, foreign invaders, enemy nations, physical phenomena and real animals. Such stories may allegorically transmit information about where it is and isn't safe to sail, tell us how different cultures believe the world was created, or contain historical narratives. The Babylonian creation epic *Enūma Eliš*, which we only know from clay tablets discovered in modern-day Iraq, is about the mixing of the primal embodiment of fresh water, Apsû, and the primal embodiment of seawater, Tiamat, and their progeny. In the story of Jonah and the Whale, the Whale is both divine punishment and allegory for the Resurrection of Christ.

Cultural groups or nations might express their maritime or naval dominance through triumphant stories of battle with serpents, or explore concerns about the dangers of seafaring through stories of ships swallowed whole. A history of sea monsters across time and space might, therefore, equate to history of the relationship between different cultures and the ocean. An examination of the sea monsters of coastal, seafaring and maritime cultures can provide us with important clues as to how these cultures relate to and conceive of the ocean and, in turn, how they have constructed themselves in accord with or in opposition to the sea depths.

But can we regard sea monsters as purely fictional entities? The Roman naturalist Pliny the Elder documented mermaids – 'tritons and nereids' – alongside dolphins and whales in his *Naturalis Historia*. In the subsection 'The sea monsters of the Indian Ocean', he describes creatures of great size, including one so large that it took the fleet of King Ptolemy 12 days to pass by. In Ovid's

RIGHT: Alexander the Great in his diving Bell, from *Histoire du bon roi Alexandre*. Unknown artist, *c.* 1300–1325.
© Courtesy of Berlin State Museums, Print Room / Jörg P. Anders

Comment Alixandres se fist avaler en la mer ou tonnel de voirre

Metaphormoses, the mortal-turned-immortal fishman Glaucus insists that he is a god, not a monster, despite his fearsome appearance.

Most people today would agree that monsters, by definition, are fictional. If a monster became known to science, it would be stripped of its monstrosity and become an animal. This is what happened to the kraken, i.e. the giant squid. A close look at the history of the kraken suggests that there is some slippage between the categories of 'monster' and 'animal' in the scientific writing of the past. The kraken has origins in Nordic mythology and folklore – in this book, *The Natural History of Norway* by the Danish bishop Erik Pontoppidan, whose description went on to influence, among others, Victor Hugo, Herman Melville and Jules Verne. By including them in his *Natural History* – an early form of scientific writing – Pontoppidan classified sea monsters as a category of animal worthy of scientific discussion just as much as fish and whales. It was not until the enlightenment that the kraken was linked to the giant squid and consequently imagined as a cephalopod. This may have been linked to the fragments of giant squid arms regurgitated by hunted sperm whales. In 1857, the Danish biologist Japetus Steenstrup first described *Architeuthis dux*, the giant squid, in the scientific journal *Forhandlinger ved de Skandinaviske Naturforskeres* (Proceedings of the Scandinavian Naturalists).

While the scientific world had thoroughly disproven the existence of a ship-sized human-eating cephalopod in the late 19th century, however, tales of sea monsters and underwater cities proliferated. From Melville's maritime novels developed a new genre of supernatural oceanic horror and early science fiction by Edgar Allan Poe, Jules Verne, HG Wells, Sir Arthur Conan Doyle, William Hope Hodgson, and HP Lovecraft. These stories reinvigorated the sea monster as a symbolic figure of contemporary fears in the age of warring empires and evolutionary anxieties. They also played upon events in existing memory: Jules Verne's *Vingt Mille Lieues Sous Les Mers* (1870) was based on an 1861 witness report from the crew of the French corvette *Alecton*, and in the novel, the *Nautilus* is suspected by many to be a giant sea monster. The kraken remained a point of fascination in society, although by then was fully relegated to the realm of fiction.

However, the kraken's shift from the realm of the mythological to the taxonomically defined realm of science reminds us that the term 'monster' is subject to cultural shift. In the medieval imagination, whales were depicted as fanged, fire-spouting beasts. During the age of whaling, whales retained their monstrousness (and are extensively referred to as monsters in *Moby-Dick*), but they were also resources to be extracted and exploited for human use. Now, in the age of mass extinction, they are beautiful and vulnerable giants in need of our protection. The mythological kraken has been taxonomically identified as the giant squid but, unlike the whale, it has not undergone the same transformation from monster to aquatic ally.

The selection of stories here seeks to interrogate the idea of the sea monster rather than arrive at any precise definition. Our subjects are tentacled, cetacean, hybrid, leviathan, chthonic, benevolent and malevolent. There are primordial entities, gods, whales, mermaids, serpents and cephalopods. There is a necessary focus on the sea monster in the English literary tradition due to the editor's breadth of expertise and the availability of English language works. There is an overwhelming bias towards male authors, as men have historically had much greater access to maritime experience. Many of the works, such as Sir Arthur Conan Doyle's novella *The Maracot Deep*, are lesser-known works of well-known writers. Others, such as *Moby-Dick* and *The Rime of the Ancient Mariner*, are highly celebrated works in their own right. The limitation to Anglophone texts is a conscious one – the texts are presented chronologically as opposed to geographically so as to trace a narrative of maritime and cultural history as sea monsters move from highly speculative, primordial entities to the apocalyptic cryptozoologist imaginings in modern science fiction in the Western imagination. *The Book of Sea Monsters*, therefore, does not claim to be an all-encompassing assessment of human civilisation's maritime mythology – what it can offer is a sweep of the sources that inform our sea monster in Western culture today by looking to the past. The progression across the book, from biblical creation monsters to the sea monsters of 20th-century genre fiction, is supported by proto-scientific and scientific writing on the natural history of sea monsters. This structure presents a genealogy and taxonomy of the sea monster, and how it *became* rather than always was fictional.

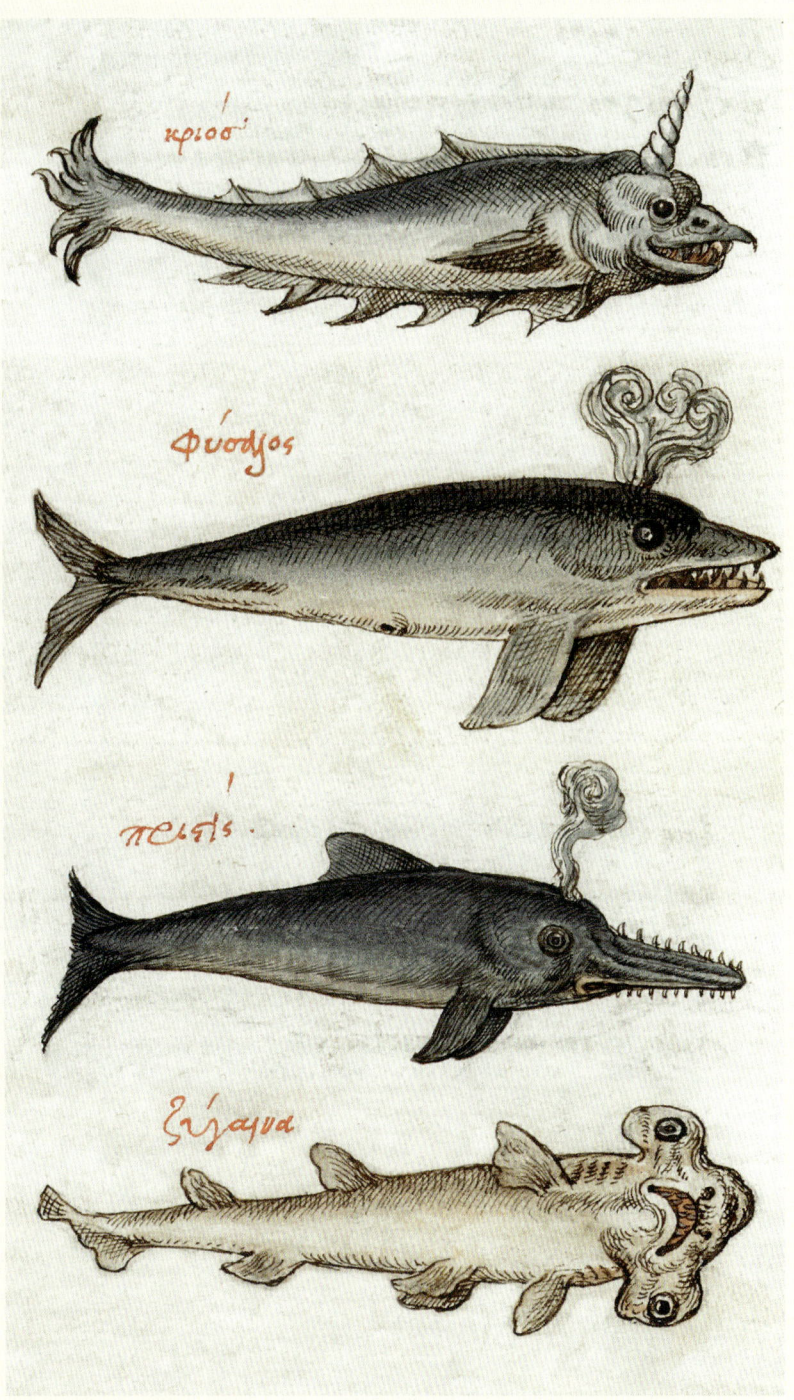

RIGHT: Drawings of fish and a hammerhead shark, from *De Animalium Proprietate*, illustrated by Angelo Vergetio, 16C.
© Courtesy of Wikimedia Commons

Introduction   9

ENŪMA ELIŠ
BY UNKNOWN
(9C BC)

The *Enūma Eliš* is the creation myth of Babylon, an ancient Mesopotamian city that existed in proximity to the Euphrates in modern-day Iraq. Our surviving sources are seven tablets in Sumero-Akkadian Cuneiform, which date back to 1200 BC. These tablets were excavated in the late 19th century at the site of Nineveh.

The first tablet describes the progeny of the two primordial beings: Apsû, the embodiment of subterranean fresh water, and Tia-mat, the sea. The mingling of their waters begets a further generation of godly children, including the four-eyed, four-eared Marduk. The noise and energetic dancing of the gods enrages Apsû, who plots to kill them but is himself slain. Taking revenge against her own progeny, Tia-mat creates 11 monsters for battle.

Tia-mat, as the primordial embodiment of seawater, is often depicted as a human woman with a tail. One of her 11 monster children, the fish-man Kulullû, also has merfolk-like features. The remaining tablets regale Marduk's triumph in battle and end with the foundation of Babylon under Marduk's instruction.

The following translation of the first tablet is from *The Babylonian Epic of Creation* by WG Lambert (1966).

LEFT: Babylonian clay map of the world, with Babylon and the Euphrates river in the centre, surrounded by a ring of ocean labelled 'bitter water'. Unknown artist/maker, *c.* 6C BC.
© Courtesy of Alamy

Enūma Eliš

When the heavens above did not exist,
And earth beneath had not come into being—
There was Apsû, the first in order, their begetter,
And demiurge Tia-mat, who gave birth to them all;
They had mingled their waters together
Before meadow-land had coalesced and reed-bed
    was to he found—
When not one of the gods had been formed
Or had come into being, when no destinies had
    been decreed,
The gods were created within them:
Lahmu and Lahamu were formed and came
    into being.
While they grew and increased in stature
Anšar and Kišar, who excelled them, were
    created.
They prolonged their days, they multiplied
    their years.
Anu, their son, could rival his fathers.
Anu, the son, equalled Anšar,
And Anu begat Nudimmud, his own equal.
Nudimmud was the champion among his fathers:
Profoundly discerning, wise, of robust strength;
Very much stronger than his father's begetter,
    Anšar
He had no rival among the gods, his brothers.
The divine brothers came together,
Their clamour got loud, throwing Tia-mat into
    a turmoil.
They jarred the nerves of Tia-mat,
And by their dancing they spread alarm
    in Anduruna.

Apsû did not diminish their clamour,
And Tia-mat was silent when confronted with them.
Their conduct was displeasing to her,
Yet though their behaviour was not good, she
    wished to spare them.
Thereupon Apsû, the begetter of the great gods,
Called Mummu, his vizier, and addressed him,
"Vizier Mummu, who gratifies my pleasure,
Come, let us go to Tia-mat!"
They went and sat, facing Tia-mat,
As they conferred about the gods, their sons.
Apsû opened his mouth
And addressed Tia-mat
"Their behaviour has become displeasing to me
And I cannot rest in the day-time or sleep
    at night.
I will destroy and break up their way of life
That silence may reign and we may sleep."
When Tia-mat heard this
She raged and cried out to her spouse,
She cried in distress, fuming within herself,
She grieved over the plotted evil,
"How can we destroy what we have given birth to?
Though their behaviour causes distress, let us
    tighten discipline graciously."
Mummu spoke up with counsel for Apsû—
(As from) a rebellious vizier was the counsel of
    his Mummu—
"Destroy, my father, that lawless way of life,
That you may rest in the day-time and sleep
    by night!"
Apsû was pleased with him, his face beamed

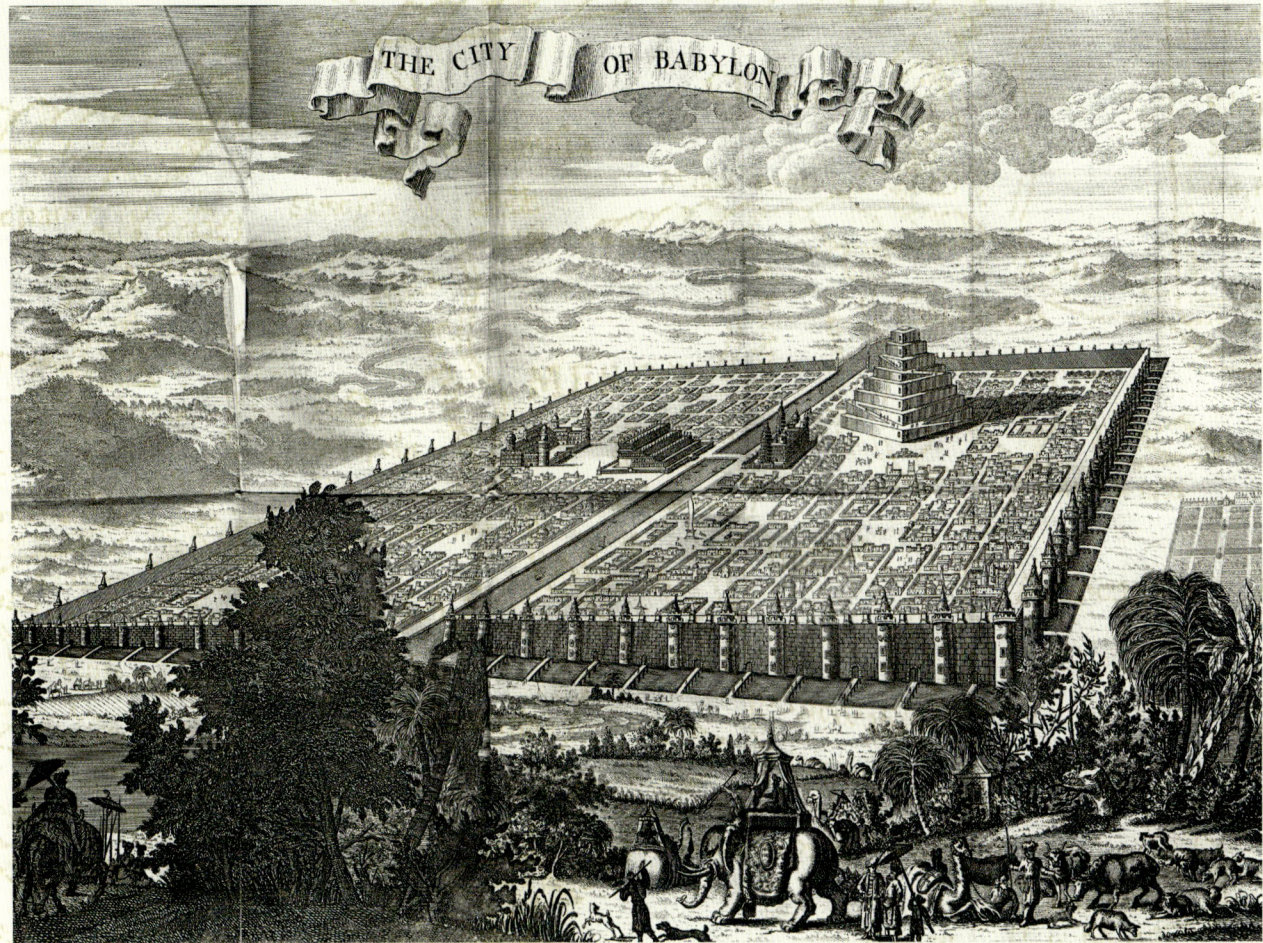

ABOVE: An illustration of the city of Babylon and surrounding area. One of the Seven Wonders of the World. 'The City of Babylon', H Fletcher, 1745.
© Courtesy of Alamy

Because he had plotted evil against the gods, his sons.
Mummu put his arms around Apsû's neck,
He sat on his knees kissing him.
What they plotted in their gathering
Was reported to the gods, their sons.
The gods heard it and were frantic.
They were overcome with silence and sat quietly.
Ea, who excels in knowledge, the skilled and learned,
Ea, who knows everything, perceived their tricks.
He fashioned it and made it to be all-embracing,
He executed it skilfully as supreme—his pure incantation.
He recited it and set it on the waters,
He poured sleep upon him as he was slumbering deeply.
He put Apsû to slumber as he poured out sleep,
And Mummu, the counsellor, was breathless with agitation.

Enūma Eliš by Unknown   13

ABOVE: Fragment of the third tablet depicting lines 47–105 of the *Enūma Eliš* in cuneiform. The tablet is made of clay and measures 2.5 by 3.5 inches (6.4 x 8.9 cm).
© Courtesy of Alamy

He split Apsû's sinews, ripped off his crown,
Carried away his aura and put it on himself.
He bound Apsû and killed him;
Mummu he confined and handled roughly.
He set his dwelling upon Apsû,
And laid hold on Mummu, keeping the nose-rope in his hand.
After Ea had bound and slain his enemies,
Had achieved victory over his foes,
He rested quietly in his chamber,
He called it Apsû, whose shrines he appointed.
Then he founded his living-quarters within it,
And Ea and Damkina, his wife, sat in splendour.
In the chamber of the destinies, the room of the archetypes,
The wisest of the wise, the sage of the gods, Be-l was conceived.
In Apsû was Marduk born,
In pure Apsû was Marduk born.
Ea his father begat him,
Damkina his mother bore him.
He sucked the breasts of goddesses,
A nurse reared him and filled him with terror.
His figure was well developed, the glance of his eyes was dazzling,
His growth was manly, he was mighty from the beginning.
Anu, his father's begeter, saw him,

14   Enūma Eliš by Unknown

He exulted and smiled; his heart filled with joy.
Anu rendered him perfect: his divinity was
 remarkable,
And he became very lofty, excelling them in
 his attributes.
His members were incomprehensibly wonderful,
Incapable of being grasped with the mind, hard
 even to look on.
Four were his eyes, four his ears,
Flame shot forth as he moved his lips.
His four ears grew large,
And his eyes likewise took in everything.
His figure was lofty and superior in comparison
 with the gods,
His limbs were surpassing, his nature was
 superior.
'Mari-utu, Mari-utu,
The Son, the Sun-god, the Sun-god of the gods.'
He was clothed with the aura of the Ten Gods, so
 exalted was his strength,
The Fifty Dreads were loaded upon him.
Anu formed and gave birth to the four winds,
He delivered them to him, "My son, let them
 whirl!"
He formed dust and set a hurricane to drive it,
He made a wave to bring consternation on
 Tia-mat.
Tia-mat was confounded; day and night she
 was frantic.
The gods took no rest, they [...]
In their minds they plotted evil,
And addressed their mother Tia-mat,
"When Apsû, your spouse, was killed,
You did not go at his side, but sat quietly.
The four dreadful winds have been fashioned
To throw you into confusion, and we cannot
 sleep.
You gave no thought to Apsû, your spouse,
Nor to Mummu, who is a prisoner. Now you
 sit alone.
Henceforth you will be in frantic consternation!
And as for us, who cannot rest, you do not love us!
Consider our burden, our eyes are hollow.
Break the immovable yoke that we may sleep.
Make battle, avenge them!
[...] reduce to nothingness!
Tia-mat heard, the speech pleased her,
She said, "Let us make demons, as you
 have advised."
The gods assembled within her.
They conceived evil against the gods
 their begetters.
They [...] and took the side of Tia-mat,
Fiercely plotting, unresting by night and day,
Lusting for battle, raging, storming,
They set up a host to bring about conflict.
Mother Hubur, who forms everything,
Supplied irresistible weapons, and gave birth to
 giant serpents.
They had sharp teeth, they were merciless [...]
With poison instead of blood she filled
 their bodies.
She clothed the fearful monsters with dread,
She loaded them with an aura and made
 them godlike.
She said, "Let their onlooker feebly perish,
May they constantly leap forward and
 never retire."
She created the Hydra, the Dragon, the Hairy Hero,
The Great Demon, the Savage Dog, and the
 Scorpion-man,
Fierce demons, the Fish-man, and the Bull-man,
Carriers of merciless weapons, fearless in the
 face of battle.

Her commands were tremendous, not to be resisted.
Altogether she made eleven of that kind.
Among the gods, her sons, whom she constituted her host,
She exalted Qingu, and magnified him among them.
The leadership of the army, the direction of the host,
The bearing of weapons, campaigning, the mobilization of conflict,
The chief executive power of battle, supreme command,
She entrusted to him and set him on a throne,
"I have cast the spell for you and exalted you in the host of the gods,
I have delivered to you the rule of all the gods.
You are indeed exalted, my spouse, you are renowned,
Let your commands prevail over all the Anunnaki."
She gave him the Tablet of Destinies and fastened it to his breast,
Saying "Your order may not be changed; let the utterance of your mouth be firm."
After Qingu was elevated and had acquired the power of Anuship,
He decreed the destinies for the gods, her sons:
"May the utterance of your mouths subdue the fire-god,
May your poison by its accumulation put down aggression."

RIGHT: 'The Fourth Map of Asia showing the eastern Mediterranean and the Middle East from Cyprus to Babylon' from *Geographia universalis*, Henricum Petrum, 1540.
© Courtesy of Wikipedia Commons

TABVLA ASIAE IIII.

ODYSSEY
BY HOMER
(8C BC)

Not much is known about the author of the *Iliad* and the *Odyssey*. By most accounts, Homer was a blind bard from Ionia who lived around the 8th century BC and travelled around ancient Greece, singing his poems with the accompaniment of a lyre. These poems were primarily orally transmitted, but at some point were transcribed.

The *Odyssey* consists of 24 books in epic verse. It takes place after the *Iliad* and chronicles the journey home of Odysseus, King of Ithaca, from the decade-long siege of Troy. The journey takes him another decade, as he is constantly waylaid by Poseidon, God of the Sea, who favoured the Trojans in the war. Odysseus blinds Poseidon's son, the cyclops Polyphemus.

The following excerpt from Book XII takes place on Scherie, the island of the Phaeacians, where Odysseus is shipwrecked once again. He tells his story to the Phaeacians, explaining the sequence of events that led him to their island, beginning with Circe's warning about navigating the waters between the six-headed sea monster Scylla and the whirlpool Charybdis. Rather than lose his whole ship to Charybdis, Odysseus sails closer to Scylla, sacrificing one man to each of her six heads. This excerpt is Samuel Butler's 1900 prose translation.

LEFT: *Herm Portrait of Blind Homer*, Unknown artist, 20C. © Digital image courtesy of Getty's Open Content Program. The J. Paul Getty Museum, Villa Collection, Malibu, California, Gift of Leon Levy, 83.AK.239

Odyssey

Book XII

"After we were clear of the river Oceanus, and had got out into the open sea, we went on till we reached the Aeaean island where there is dawn and sunrise as in other places. We then drew our ship on to the sands and got out of her on to the shore, where we went to sleep and waited till day should break.

"Then, when the child of morning, rosy-fingered Dawn, appeared, I sent some men to Circe's house to fetch the body of Elpenor. We cut firewood from a wood where the headland jutted out into the sea, and after we had wept over him and lamented him we performed his funeral rites. When his body and armour had been burned to ashes, we raised a cairn, set a stone over it, and at the top of the cairn we fixed the oar that he had been used to row with.

"While we were doing all this, Circe, who knew that we had got back from the house of Hades, dressed herself and came to us as fast as she could; and her maid servants came with her bringing us bread, meat, and wine. Then she stood in the midst of us and said, 'You have done a bold thing in going down alive to the house of Hades, and you will have died twice, to other people's once; now, then, stay here for the rest of the day, feast your fill, and go on with your voyage at daybreak tomorrow morning. In the meantime I will tell Ulysses about your course, and will explain everything to him so as to prevent your suffering from misadventure either by land or sea.'

"We agreed to do as she had said, and feasted through the livelong day to the going down of the sun, but when the sun had set and it came on dark, the men laid themselves down to sleep by the stern cables of the ship. Then Circe took me by the hand and bade me be seated away from the others, while she reclined by my side and asked me all about our adventures.

"'So far so good,' said she, when I had ended my story, 'and now pay attention to what I am about to tell you—heaven itself, indeed, will recall it to your recollection. First you will come to the Sirens who enchant all who come near them. If any one unwarily draws in too close and hears the singing of the Sirens, his wife and children will never welcome him home again, for they sit in a green field and warble him to death with the sweetness of their song. There is a great heap of dead men's bones lying all around, with the flesh still rotting off them. Therefore pass these Sirens by, and stop your men's ears with wax that none of them may hear; but if you like you can listen yourself, for you may get the men to bind you as you stand upright on a cross piece half way up the mast, and they must lash the rope's ends to the mast itself, that you may have the pleasure of listening. If you beg and pray the men to unloose you, then they must bind you faster.

"'When your crew have taken you past these Sirens, I cannot give you coherent directions as to which of two courses you are to take; I will lay the two alternatives before you, and you must consider them for yourself. On the one hand there are some overhanging rocks against which the deep blue waves of Amphitrite beat with terrific fury; the blessed gods call these rocks the Wanderers. Here not even a bird may pass, no, not even the timid doves that bring ambrosia to Father Jove, but the sheer rock always carries off one of them, and Father Jove has to send another to make up their number; no ship that ever yet came to these rocks has got away again, but the waves and whirlwinds of fire are freighted with wreckage and with the bodies of dead men. The only vessel that ever sailed and got through, was the famous Argo on her way from the house of Aetes, and she too would have gone against these great rocks, only that Juno piloted her past them for the love she bore to Jason.

ABOVE: Scylla (left) and Sirens (right). Unknown artist/maker/illuminator, c. 1475.
© Digital image courtesy of Getty's Open Content Program. The J. Paul Getty Museum, Los Angeles, Ms. Ludwig XIII 5, v1, fol. 68v, 83.MP.148.1.68v

Odyssey by Homer

"'Of these two rocks the one reaches heaven and its peak is lost in a dark cloud. This never leaves it, so that the top is never clear not even in summer and early autumn. No man though he had twenty hands and twenty feet could get a foothold on it and climb it, for it runs sheer up, as smooth as though it had been polished. In the middle of it there is a large cavern, looking West and turned towards Erebus; you must take your ship this way, but the cave is so high up that not even the stoutest archer could send an arrow into it. Inside it Scylla sits and yelps with a voice that you might take to be that of a young hound, but in truth she is a dreadful monster and no one—not even a god—could face her without being terror-struck. She has twelve mis-shapen feet, and six necks of the most prodigious length; and at the end of each neck she has a frightful head with three rows of teeth in each, all set very close together, so that they would crunch any one to death in a moment, and she sits deep within her shady cell thrusting out her heads and peering all round the rock, fishing for dolphins or dogfish or any larger monster that she can catch, of the thousands with which Amphitrite teems. No ship ever yet got past her without losing some men, for she shoots out all her heads at once, and carries off a man in each mouth.

"'You will find the other rocks lie lower, but they are so close together that there is not more than a bow-shot between them. A large fig tree in full leaf grows upon it, and under it lies the sucking whirlpool of Charybdis. Three times in the day does she vomit forth her waters, and three times she sucks them down again; see that you be not there when she is sucking, for if you are, Neptune himself could not save you; you must hug the Scylla side and drive ship by as fast as you can, for you had better lose six men than your whole crew.'

"'Is there no way,' said I, 'of escaping Charybdis, and at the same time keeping Scylla off when she is trying to harm my men?'

"'You dare devil,' replied the goddess, 'you are always wanting to fight somebody or something; you will not let yourself be beaten even by the immortals. For Scylla is not mortal; moreover she is savage, extreme, rude, cruel and invincible. There is no help for it; your best chance will be to get by her as fast as ever you can, for if you dawdle about her rock while you are putting on your armour, she may catch you with a second cast of her six heads, and snap up another half dozen of your men; so drive your ship past her at full speed, and roar out lustily to Crataiis who is Scylla's dam, bad luck to her; she will then stop her from making a second raid upon you.'

"'You will now come to the Thrinacian island, and here you will see many herds of cattle and flocks of sheep belonging to the sun-god—seven herds of cattle and seven flocks of sheep, with fifty head in each flock. They do not breed, nor do they become fewer in number, and they are tended by the goddesses Phaethusa and Lampetie, who are children of the sun-god Hyperion by Neaera. Their mother when she had borne them and had done suckling them sent them to the Thrinacian island, which was a long way off, to live there and look after their father's flocks and herds. If you leave these flocks unharmed, and think of nothing but getting home, you

ABOVE: A Siren lures a sailor, *Dicta Chrysostomi*, 1280–1300.
© Courtesy of Wikimedia Commons

may yet after much hardship reach Ithaca; but if you harm them, then I forewarn you of the destruction both of your ship and of your comrades; and even though you may yourself escape, you will return late, in bad plight, after losing all your men.'

"Here she ended, and dawn enthroned in gold began to show in heaven, whereon she returned inland. I then went on board and told my men to loose the ship from her moorings; so they at once got into her, took their places, and began to smite the grey sea with their oars. Presently the great and cunning goddess Circe befriended us with a fair wind that blew dead aft, and staid steadily with us, keeping our sails well filled, so we did whatever wanted doing to the ship's gear, and let her go as wind and helmsman headed her.

"Then, being much troubled in mind, I said to my men, 'My friends, it is not right that one or two of us alone should know the prophecies that Circe has made me, I will therefore tell you about them, so that whether we live or die we may do so with our eyes open. First she said

Odyssey by Homer   23

we were to keep clear of the Sirens, who sit and sing most beautifully in a field of flowers; but she said I might hear them myself so long as no one else did. Therefore, take me and bind me to the crosspiece half way up the mast; bind me as I stand upright, with a bond so fast that I cannot possibly break away, and lash the rope's ends to the mast itself. If I beg and pray you to set me free, then bind me more tightly still.'

"I had hardly finished telling everything to the men before we reached the island of the two Sirens, for the wind had been very favourable. Then all of a sudden it fell dead calm; there was not a breath of wind nor a ripple upon the water, so the men furled the sails and stowed them; then taking to their oars they whitened the water with the foam they raised in rowing. Meanwhile I look a large wheel of wax and cut it up small with my sword. Then I kneaded the wax in my strong hands till it became soft, which it soon did between the kneading and the rays of the sun-god son of Hyperion. Then I stopped the ears of all my men, and they bound me hands and feet to the mast as I stood upright on the cross piece; but they went on rowing themselves. When we had got within earshot of the land, and the ship was going at a good rate, the Sirens saw that we were getting in shore and began with their singing.

"'Come here,' they sang, 'renowned Ulysses, honour to the Achaean name, and listen to our two voices. No one ever sailed past us without staying to hear the enchanting sweetness of our song—and he who listens will go on his way not only charmed, but wiser, for we know all the ills that the gods laid upon the Argives and Trojans before Troy, and can tell you everything that is going to happen over the whole world.'

"They sang these words most musically, and as I longed to hear them further I made signs by frowning to my men that they should set me free; but they quickened their stroke, and Eurylochus and Perimedes bound me with still stronger bonds till we had got out of hearing of the Sirens' voices. Then my men took the wax from their ears and unbound me.

"Immediately after we had got past the island I saw a great wave from which spray was rising, and I heard a loud roaring sound. The men were so frightened that they loosed hold of their oars, for the whole sea resounded with the rushing of the waters, but the ship stayed where it was, for the men had left off rowing. I went round, therefore, and exhorted them man by man not to lose heart.

"'My friends,' said I, 'this is not the first time that we have been in danger, and we are in nothing like so bad a case as when the Cyclops shut us up in his cave; nevertheless, my courage and wise counsel saved us then, and we shall live to look back on all this as well. Now, therefore, let us all do as I say, trust in Jove and row on with might and main. As for you, coxswain, these are your orders; attend to them, for the ship is in your hands; turn her head away from these steaming rapids and hug the rock, or she will give you the slip and be over yonder before you know where you are, and you will be the death of us.'

"So they did as I told them; but I said nothing about the awful monster Scylla, for I knew the men would not go on rowing if I did, but would huddle together in the hold. In one thing only did I disobey Circe's strict instructions—I put on my armour. Then seizing two strong spears I took my stand on the ship's bows, for it was there that I expected first to see the monster of the rock, who was to do my men so much harm; but I could not make her out anywhere, though I strained my eyes with looking the gloomy rock all over and over.

"Then we entered the Straits in great fear of mind, for on the one hand was Scylla, and on the other dread Charybdis kept sucking up the salt water. As she vomited it up, it was like the water in a cauldron when it is boiling over upon a great fire, and the spray reached the top of the rocks on either side. When she began to suck again, we could see the water all inside whirling round and round, and it made a deafening sound as it broke against the rocks. We could see the bottom of the whirlpool all black with sand and mud, and the men were at their wits ends for fear. While we were taken up with this, and were expecting each moment to be our last, Scylla pounced down suddenly upon us and snatched up my six best men. I was looking at once after both ship and men, and in a moment I saw their hands and feet ever so high above me, struggling in the air as Scylla was carrying them off, and I heard them call out my name in one last despairing cry. As a fisherman, seated, spear in hand, upon some jutting rock throws bait into the water to deceive the poor little fishes, and spears them with the ox's horn with which his spear is shod, throwing them gasping on to the land as he catches them one by one—even so did Scylla land these panting creatures on her rock and munch them up at the mouth of her den, while they screamed and stretched out their hands to me in their mortal agony. This was the most sickening sight that I saw throughout all my voyages.

PREVIOUS PAGES: *Charybde et Scylla*, Ary Ernest Renan, 1894.
© Courtesy of Musée de la Vie romantique, CSR P 17

THE AENEID
BY VERGIL
(19C BC)

Publius Vergilius Maro or Vergil (sometimes spelled Virgil) (70–19 BC) was a Roman poet during the reign of Caesar Augustus, who founded the Roman Empire in 27 BC. Vergil is the author of three great works: the *Eclogues*, a long bucolic poem; the *Georgics*, an agricultural poem and political eulogy; and the *Aeneid*, an epic poem about the journey of the Trojan prince Aeneas.

The *Aeneid* was commissioned by Augustus and took Vergil 11 years to write. It consists of 12 books in dactylic hexameter, adapting the meter of Greek heroic verse into Latin. The *Aeneid* is a work of political propaganda, justifying and celebrating Augustus's rule over Rome, as Aeneas is told that his descendant Romulus will eventually be the founder of the city.

The following excerpt takes place as Aeneas narrates to Dido, Queen of Carthage, about how the Greeks infiltrated Troy with a wooden horse. Laocoön, a Trojan priest of Neptune, throws a spear into the horse to show that it is hollow. In retaliation, Minerva (the Roman equivalent of Athena) sends two sea-serpents to devour Laocoön and his two sons.

This version of the *Aeneid* was translated by Theodore C Williams for Houghton Mifflin Co in 1910.

LEFT: Laocoön, from 'Speculum Romanae Magnificentiae' (The Mirror of Roman Magnificence), Nicolas Beatrizet (artist), Antonio Lafreri (publisher), 6C.
© Courtesy of The Met Museum, Harris Brisbane Dick Fund, 1941, 41.72(2.76)

The Aeneid

Book II, ll. 200–233

But now a vaster spectacle of fear
burst over us, to vex our startled souls.
Laocoön, that day by cast of lot
priest unto Neptune, was in act to slay
a huge bull at the god's appointed fane.
Lo! o'er the tranquil deep from Tenedos
appeared a pair (I shudder as I tell)
of vastly coiling serpents, side by side,
stretching along the waves, and to the shore
taking swift course; their necks were lifted high,
their gory dragon-crests o'ertopped the waves;
all else, half seen, trailed low along the sea;
while with loud cleavage of the foaming brine
their monstrous backs wound forward fold on fold.
Soon they made land; the furious bright eyes
glowed with ensanguined fire; their
    quivering tongues
lapped hungrily the hissing, gruesome jaws.
All terror-pale we fled. Unswerving then
the monsters to Laocoön made way.
First round the tender limbs of his two sons
each dragon coiled, and on the shrinking flesh
fixed fast and fed. Then seized they on the sire,
who flew to aid, a javelin in his hand,
embracing close in bondage serpentine
twice round the waist; and twice in scaly grasp
around his neck, and o'er him grimly peered
with lifted head and crest; he, all the while,
his holy fillet fouled with venomous blood,
tore at his fetters with a desperate hand,
and lifted up such agonizing voice,
as when a bull, death-wounded, seeks to flee
the sacrificial altar, and thrusts back
from his doomed head the ill-aimed, glancing blade.
then swiftly writhed the dragon-pair away
unto the templed height, and in the shrine
of cruel Pallas sure asylum found
beneath the goddess' feet and orbed shield.
Such trembling horror as we ne'er had known
seized now on every heart. "Of his vast guilt
Laocoön," they say, "receives reward;
for he with most abominable spear
did strike and violate that blessed wood.
Yon statue to the temple! Ask the grace
of glorious Pallas!" So the people cried
in general acclaim.

ABOVE: *Virgil Reading the 'Aeneid' to Augustus, Octavia and Livia*, Jean Baptiste Joseph Wicar, 1790.
© Courtesy of The Art Institute Chicago, Wirt D. Walker Fund, 1963.258, Creative Commons Zero (CC0)

METAMORPHOSES
BY OVID
(AD 8)

Publius Ovidius Naso (43 BC–AD 17 or 18) was a Roman poet during the reign of Caesar Augustus, succeeding Vergil. Ovid travelled across the Empire and lived most of his life in Rome until AD 8, when Octavian exiled him to Tomis in modern-day Romania for reasons Ovid himself explained as *carmen et error*: 'a poem and a mistake'.

Ovid's most well-known work is the *Metamorphoses*, written in the year AD 8. It is a poem recounting stories of transformation in Greco-Roman mythology. It consists of 25 books, each containing multiple fables told in a continuous style with each fable following after the last.

The following prose translation is from Henry T Riley's 1893 *The Metamorphoses of Ovid* Book XIII, Fable VIII, on Glaucus, and Book XIV, Fable I on Scylla. Glaucus, according to Ovid, is a mortal who observed dead fish coming to life after eating a magical herb. Upon trying this herb himself, he gains immortality and throws himself into the sea, transforming into a green-bearded, azure-hued fishman with a tail. Scylla is best known from her presence in Homer's *Odyssey* (see page 19), in which she is a six-headed monster. Ovid tells the story of the nymph Scylla's transformation, first into a hideous monster by Circe, then into a rock as she throws herself into the sea.

LEFT: *Scylla and Glaucus*, etching by Antonio Tempesta, 1606.
© Courtesy of The Met Museum, 51.501.3981

Metamorphoses

Book XIII, Fable VIII (Glaucus)

Galatea ceases speaking, and the company breaking up, they depart; and the Nereids swim in the becalmed waves. Scylla returns, (for, in truth, she does not trust herself in the midst of the ocean) and either wanders about without garments on the thirsty sand, or, when she is tired, having lighted upon some lonely recess of the sea, cools her limbs in the enclosed waves. When, lo! cleaving the deep, Glaucus comes, a new-made inhabitant of the deep sea, his limbs having been lately transformed at Anthedon, near Eubœa; and he lingers from passion for the maiden now seen, and utters whatever words he thinks may detain her as she flies. Yet still she flies, and, swift through fear, she arrives at the top of a mountain, situate near the shore.

In front of the sea, there is a huge ridge, terminating in one summit, bending for a long distance over the waves, and without trees. Here she stands, and secured by the place, ignorant whether he is a monster or a God, she both admires his colour, and his flowing hair that covers his shoulders and his back, and how a wreathed fish closes the extremity of his groin. This he perceives; and leaning upon a rock that stands hard by, he says,

'Maiden, I am no monster, no savage beast; I am a God of the waters: nor have Proteus, and Triton, and Palæmon, the son of Athamas, a more uncontrolled reign over the deep. Yet formerly I was a mortal; but, still, devoted to the deep sea, even then was I employed in it. For, at one time, I used to drag the nets that swept up the fish; at another time, seated on a rock, I managed the line with the rod. The shore was adjacent to a verdant meadow, one part of which was surrounded with water, the other with grass, which, neither the horned heifers had hurt with their browsing, nor had you, ye harmless sheep, nor you, ye shaggy goats, ever cropped it. No industrious bee took thence the collected blossoms, no festive garlands were gathered thence for the head; and no mower's hands had ever cut it. I was the first to be seated on that turf, while I was drying the dripping nets. And that I might count in their order the fish that I had taken; I laid out those upon it which either chance had driven to my nets, or their own credulity to my barbed hooks.

'The thing is like a fiction (but of what use is it to me to coin fictions?); on touching the grass my prey began to move, and to shift their sides, and to skip about on the land, as though in the sea. And while I both paused and wondered, the whole batch flew off to the waves, and left behind their new master and the shore. I was amazed, and, in doubt for a long time, I considered what could be the cause; whether some Divinity had done this, or whether the juice of some herb. 'And yet,' said I, 'what herb has these properties?' and with my hand I plucked the grass, and I chewed it, so plucked, with my teeth. Hardly had my throat well swallowed the unknown juices, when I suddenly felt my entrails

RIGHT: *Glaucus and Scylla*, Bartholomeus Spranger, 1582.
© Courtesy of Wikimedia Commons

inwardly throb, and my mind taken possession of by the passions of another nature. Nor could I stay in that place; and I exclaimed, 'Farewell, land, never more to be revisited;' and plunged my body beneath the deep. The Gods of the sea vouchsafed me, on being received by them, kindred honours, and they entreated Oceanus and Tethys to take away from me whatever mortality I bore. By them was I purified; and a charm being repeated over me nine times, that washes away all guilt, I was commanded to put my breast beneath a hundred streams.

'There was no delay; rivers issuing from different springs, and whole seas, were poured over my head. Thus far I can relate to thee what happened worthy to be related, and thus far do I remember; but my understanding was not conscious of the rest. When it returned to me, I found myself different throughout all my body from what I was before, and not the same in mind. Then, for the first time, did I behold this beard, green with its deep colour, and my flowing hair, which I sweep along the spacious seas, and my huge shoulders, and my azurecoloured arms, and the extremities of my legs tapering in the form of a finny fish. But still, what does this form avail me, what to have pleased the ocean Deities, and what to be a God, if thou art not moved by these things?'

As he was saying such things as these, and about to say still more, Scylla left the God. He was enraged, and, provoked at the repulse, he repaired to the marvellous court of Circe, the daughter of Titan.

Book XIV, Fable I (Scylla)

And now Glaucus, the Eubœan plougher of the swelling waves, had left behind Ætna, placed upon the jaws of the Giant, and the fields of the Cyclops, that had never experienced the harrow or the use of the plough, and that were never indebted to the yoked oxen; he had left Zancle, too, behind, and the opposite walls of Rhegium, and the sea, abundant cause of shipwreck, which, confined by the two shores, bounds the Ausonian and the Sicilian lands. Thence, swimming with his huge hands through the Etrurian seas, Glaucus arrived at the grass-clad hills, and the halls of Circe, the daughter of the Sun, filled with various wild beasts. Soon as he beheld her, after salutations were given and received, he said,

'Do thou, a Goddess, have compassion on me a God; for thou alone (should I only seem deserving of it,) art able to relieve this passion of mine. Daughter of Titan, by none is it better known how great is the power of herbs, than by me, who have been transformed by their agency; and, that the cause of my passion may not be unknown to thee, Scylla has been beheld by me on the Italian shores, opposite the Messenian walls. I am ashamed to recount my promises, my entreaties, my caresses, and my rejected suit. But, do thou, if there is any power in incantations, utter the incantation with thy holy lips; or, if any herb is more efficacious, make use of the proved virtues of powerful herbs. But I do not request thee to cure me, and to heal these wounds; and there is no necessity for an end to them; but let her share in the flame.'

But Circe, (for no one has a temper more susceptible of such a passion, whether it is that the cause of it originates in herself, or whether it is that Venus, offended by her father's discovery, causes this,) utters such words as these:—

'Thou wilt more successfully court her who is willing, and who entertains similar desires, and who is captivated with an equal passion. Thou art worthy of it, and assuredly thou oughtst to be courted spontaneously; and, if thou givest any hopes, believe me, thou shalt be courted spontaneously. That thou mayst entertain no doubts, or lest confidence in thy own beauty may not exist, behold! I who am both a Goddess, and the daughter of the radiant Sun, and am so potent with my charms, and so potent with my herbs, wish to be thine. Despise her who despises thee; her, who is attached to thee, repay by like attachment, and, by one act, take vengeance on two individuals.'

Glaucus answered her, making such attempts as these:—

'Sooner shall foliage grow in the ocean, and sooner shall sea-weed spring up on the tops of the mountains, than my affections shall change, while Scylla is alive.'

The Goddess is indignant; and since she is not able to injure him, and as she loves him she does not wish to do so, she is enraged against her, who has been preferred to herself; and, offended with these crosses in love, she immediately bruises herbs, infamous for their horrid juices, and, when bruised, she mingles with them the incantations of Hecate. She puts on azure vestments too, and through the troop of fawning wild beasts she issues from the midst of her hall; and making for Rhegium, opposite to the rocks of Zancle, she enters the waves boiling with the tides; on these, as though on the firm shore, she impresses her footsteps, and with dry feet she skims along the surface of the waves.

There was a little bay, curving in the shape of a bent bow, a favourite retreat of Scylla, whither she used to retire from the influence both of the sea and of the weather, when the sun was at its height in his mid career, and made the smallest shadow from the head downwards. This the Goddess infects beforehand, and pollutes it with monster-breeding drugs; on it she sprinkles the juices distilled from the noxious root, and thrice nine times, with her magic lips, she mutters over the mysterious charm, enwrapt in the dubious language of strange words. Scylla comes; and she has now gone in up to the middle of her stomach, when she beholds her loins grow hideous with barking monsters; and, at first believing that they are no part of her own body, she flies from them and drives them off, and is in dread of the annoying mouths of the dogs; but those that she flies from, she carries along with herself; and as she examines the substance of her thighs, her legs, and her feet, she meets with Cerberean jaws in place of those parts. The fury of the dogs still continues, and the backs of savage monsters lying beneath her groin, cut short, and her prominent stomach, still adhere to them.

Glaucus, still in love, bewailed her, and fled from an alliance with Circe, who had thus too hostilely employed the potency of herbs. Scylla remained on that spot; and, at the first moment that an opportunity was given, in her hatred of Circe, she deprived Ulysses of his companions. Soon after, the same Scylla would have overwhelmed the Trojan ships, had she not been first transformed into a rock, which even now is prominent with its crags; this rock the sailor, too, avoids.

NATURALIS HISTORIA
BY PLINY THE ELDER
(AD 77)

Pliny the Elder or Gaius Plinius Secundus (AD 23–79) was a writer and natural philosopher of the early Roman Empire. He died in an attempt to rescue his friend Rectina from the eruption of Mount Vesuvius that destroyed Pompeii.

The 36-book *Naturalis Historia* (Natural History), first published in AD 77, is Pliny's only surviving work. It is, according to Pliny's dedication, an account of '20,000 topics, all worthy of attention […] gained by the perusal of about 2,000 volumes […] and to these I have made considerable additions of things, which were either not known to my predecessors, or which have been lately discovered'. Pliny covers all aspects of the natural world, with books dedicated to the elements, landforms, animals, humans, botany, medicine, metals and stones. Its breadth and structure influenced the model of the encyclopaedia as well as defined the scope of natural history as a pursuit. Drawing upon an enormous range of sources in addition to personal experience, he describes all aspects of the natural world from astronomy, agriculture and zoology to horticulture and geology.

The following excerpts are from Book IX of the *Naturalis Historia*, 'The Natural History of Fishes'. In this book, Pliny describes all fishes known to Rome. Tritons and nereids (mermaids) are described alongside whales and dolphins, as well as fish that contemporary scholars have been unable to equate with known species.

LEFT: Pliny the Elder. © Courtesy of *Encyclopaedia Britannica*, Mary Evans Picture Library Ltd/age fotostock

Naturalis Historia Book IX: The Natural History of Fishes

'Why the largest animals are found in the sea.'

We have now given an account of the animals which we call terrestrial, and which live as it were in a sort of society with man. Among the remaining ones, it is well known that the birds are the smallest; we shall therefore first describe those which inhabit the seas, rivers, and standing waters.

Among these there are many to be found that exceed in size any of the terrestrial animals even; the evident cause of which is the superabundance of moisture with which they are supplied. Very different is the lot of the winged animals, whose life is passed soaring aloft in the air. But in the seas, spread out as they are far and wide, forming an element at once so delicate and so vivifying, and receiving the generating principles from the regions of the air, as they are ever produced by Nature, many animals are to be found, and indeed, most of those that are of monstrous form; from the fact, no doubt, that these seeds and first principles of being are so utterly conglomerated and so involved, the one with the other, from being whirled to and fro, now by the action of the winds and now by the waves. Hence it is that the vulgar notion may very possibly be true, that whatever is produced in any other department of Nature, is to be found in the sea as well; while, at the same time, many other productions are there to be found which nowhere else exist. That there are to be found in the sea the forms, not only of terrestrial animals, but of inanimate objects even, is easily to be understood by all who will take the trouble to examine the grape-fish, the sword-fish, the sawfish, and the cucumber-fish, which last so strongly resembles the real cucumber both in colour and in smell. We shall find the less reason then to be surprised to find that in so small an object as a shell-fish the head of the horse is to be seen protruding from the shell.

'The sea monsters of the Indian Ocean.'

But the most numerous and largest of all these animals are those found in the Indian seas; among which there are balænæ, four jugera in extent, and the pristis, two hundred cubits long: here also are found cray-fish four cubits in length, and in the river Ganges there are to be seen eels three hundred feet long. But at sea it is more especially about the time of the

ABOVE: 'Polyp' (octopus) from the *Hortus Sanitatis*, Jacobus Meydenbach, 1491. © Courtesy of the Wellcome Collection

40   Naturalis Historia by Pliny the Elder

ABOVE: Sea monsters attacking a ship, *Indiae Orientalis*, Abraham Ortelius, 1570.
© Courtesy of Wikimedia Commons

Naturalis Historia by Pliny the Elder 41

ABOVE: Sea monster illustration from *Islandia*, Abraham Ortelius, 1585.
© Courtesy of Wikimedia Commons

solstices that these monsters are to be seen. For then it is that in these regions the whirlwind comes sweeping on, the rains descend, the hurricane comes rushing down, hurled from the mountain heights, while the sea is stirred up from the very bottom, and the monsters are driven from their depths and rolled upwards on the crest of the billow. At other times again, there are such vast multitudes of tunnies met with, that the fleet of Alexander the Great was able to make head against them only by facing them in order of battle, just as it would have done an enemy's fleet. Had the ships not done this, but proceeded in a straggling manner, they could not possibly have made their escape. No noises, no sounds, no blows had any effect on these fish; by nothing short of the clash of battle were they to be terrified, and by nothing less than their utter destruction were they overpowered.

There is a large peninsula in the Red Sea, known by the name of Cadara as it projects into the deep it forms a vast gulf, which it took the fleet of King Ptolemy twelve whole days and nights to traverse by dint of rowing, for not a breath of wind was to be perceived. In the recesses of this becalmed spot more particularly, the sea-monsters attain so vast a size that they are quite unable to move. The commanders of the fleets of Alexander the Great have related that the Gedrosi, who dwell upon the banks of the river Arabis, are in the habit of making the doors of their houses with the jaw-bones of fishes, and raftering the roofs with their bones, many of which were found as much as forty cubits in length. At this place, too, the sea-monsters, just like so many cattle, were in the habit of coming on shore, and, after feeding on the roots of shrubs, they would return; some of them, which had the heads of horses, asses, and bulls, found a pasture in the crops of grain.

'The largest animals that are found in each ocean.'

The largest animals found in the Indian Sea are the pistrix and the balæna; while of the Gallic Ocean the physeter is the most bulky inhabitant, raising itself aloft like some vast column, and as it towers above the sails of ships, belching forth, as it were, a deluge of water. In the ocean of Gades there is a tree, with outspread branches so vast, that it is supposed that it is for that reason it has never yet entered the Straits. There are fish also found there which are called sea-wheels, in consequence of their singular conformation; they are divided by four spokes, the nave being guarded on every side by a couple of eyes.

ABOVE: Merman playing an instrument, *Septentrionalium regionum descrip*, Abraham Ortelius, 1570.
© Courtesy of Wikimedia Commons

'The forms of the tritons and nereids. The forms of sea elephants.'

A deputation of persons from Olisipo, that had been sent for the purpose, brought word to the Emperor Tiberius that a triton had been both seen and heard in a certain cavern, blowing a conch-shell, and of the form under which they are usually represented. Nor yet is the figure generally attributed to the nereids at all a fiction; only in them, the portion of the body that resembles the human figure is still rough all over with scales. For one of these creatures was seen upon the same shores, and as it died, its plaintive murmurs were heard even by the inhabitants at a distance. The legatus of Gaul, too, wrote word to the late Emperor Augustus that a considerable number of nereids had been found dead upon the sea-shore. I have, too, some distinguished informants of equestrian rank, who state that they themselves once saw in the ocean of Gades a sea-man, which bore in every part of his body a perfect resemblance to a human being, and that during the night he would climb up into ships; upon which the side of the vessel where he seated himself would instantly sink downward, and if he remained there any considerable time, even go under water.

In the reign of the Emperor Tiberius, a subsidence of the ocean left exposed on the shores of an island which faces the province of Lugdunum as many as three hundred animals or more, all at once, quite marvellous for their varied shapes and enormous size, and no less a number upon the shores of the Santones; among the rest there were elephants and rams, which last, however, had only a white spot to represent horns. Turranius has also left accounts of several nereids, and he speaks of a monster that was thrown up on the shore at Gades, the distance between the two fins at the end of the tail of which was sixteen cubits, and its teeth one hundred and twenty in number; the largest being nine, and the smallest six inches in length.

M. Scaurus, in his ædileship, exhibited at Rome, among other wonderful things, the bones of the monster to which Andromeda was said to have been exposed, and which he had brought from Joppa, a city of Judæa. These bones exceeded forty feet in length, and the ribs were higher than those of the Indian elephant, while the back-bone was a foot and a half in thickness.

'The balæna and the orca'

The balæna penetrates to our seas even. It is said that they are not to be seen in the ocean of Gades before the winter solstice, and that at periodical seasons they retire and conceal themselves in some calm capacious bay, in which they take a delight in bringing forth. This fact, however, is known to the orca, an animal which is peculiarly hostile to the balæna, and the form of which cannot be in any way adequately described, but as an enormous mass of flesh armed with teeth. This animal attacks the balæna in its places of retirement, and with its teeth tears its young, or else attacks the females which have just brought forth, and, indeed, while they are still pregnant: and as they rush upon them, it pierces them just as though they had been attacked by the beak of a Liburnian galley. The female balænæ, devoid of all flexibility, without energy to defend themselves, and over-burdened by their own weight, weakened, too, by gestation, or else the pains of recent parturition, are well aware that their only resource is to take to flight in the open sea and to range over the whole face of the ocean; while the orcæ, on the other hand, do all in their power to meet them in their flight, throw themselves in their way, and kill them either cooped up in a narrow passage, or else drive them on a shoal, or dash them to pieces against the rocks. When these battles are witnessed, it appears just as though the sea were infuriate against itself; not a breath of wind is there to be felt in the bay, and yet the waves by their pantings and their repeated blows are heaved aloft in a way which no whirlwind could effect.

An orca has been seen even in the port of Ostia, where it was attacked by the Emperor Claudius. It was while he was

ABOVE: Balena or 'Whale'. *De aquatilibus*, Apud C Stephanum, 1553.
© Courtesy of the Biodiversity Heritage Library

constructing the harbour there that this orca came, attracted by some hides which, having been brought from Gaul, had happened to fall overboard there. By feeding upon these for several days it had quite glutted itself, having made for itself a, channel in the shoaly water. Here, however, the sand was thrown up by the action of the wind to such an extent, that the creature found it quite impossible to turn round; and while in the act of pursuing its prey, it was propelled by the waves towards the shore, so that its back came to be perceived above the level of the water, very much resembling in appearance the keel of a vessel turned bottom upwards. Upon this, Cæsar ordered a great number of nets to be extended at the mouth of the harbour, from shore to shore, while he himself went there with the prætorian cohorts, and so afforded a spectacle to the Roman people; for boats assailed the monster, while the soldiers on board showered lances upon it. I myself saw one of the boats sunk by the water which the animal, as it respired, showered down upon it.

Naturalis Historia by Pliny the Elder

THE BOOK OF JONAH

BY UNKNOWN
(DATE UNKNOWN)

The Book of Jonah first appears in the Tanach, the Hebrew Bible. It tells the story of the prophet Jonah, who defies God's orders to go to Nineveh and instead flees on a ship to Tarshish. God creates a storm in retaliation, and the sailors cast Jonah overboard. He is swallowed by a 'great fish' or 'whale', where he spends three days before God releases him.

In Christian typology (the exegetical practice of reading the Old Testament as prefiguring the events of the New Testament), Jonah's three days spent inside the whale are interpreted as allegorical for the resurrection of Christ, who also spent three days entombed. The Book of Jonah is typically read on the Jewish holiday of Yom Kippur, the Day of Atonement.

In Islam, Jonah is also considered a prophet by the name of Yūnus ibn Mattā. The 10th chapter of the Quran, Surah Yūnus (The Book of Jonah), is named after him, but his encounter with the whale is referenced in the 37th chapter, Surah al-Saffat, The Book of the Rangers.

Although we generally refer to the story of Jonah and the Whale, the Hebrew דָּג גָּדוֹל, *dāḡ gāḏōl* meaning 'great fish', translates to the Greek *kētos* and Latin *cētus*, which themselves have been translated as 'sea monster', 'great fish' or 'whale'. The following excerpt from the first half of the Book of Jonah is taken from the 1611 King James Bible (KJV), which uses 'great fish'.

LEFT: Jonah preaching to the Ninevites. *La Grande Bible de Tours*, Gustave Doré, 1866.
© Courtesy of Wikimedia Commons

The Book of Jonah

Jonah 1

Now the word of the Lord came unto Jonah the son of Amittai, saying,

Arise, go to Nineveh, that great city, and cry against it; for their wickedness is come up before me.

But Jonah rose up to flee unto Tarshish from the presence of the Lord, and went down to Joppa; and he found a ship going to Tarshish: so he paid the fare thereof, and went down into it, to go with them unto Tarshish from the presence of the Lord.

But the Lord sent out a great wind into the sea, and there was a mighty tempest in the sea, so that the ship was like to be broken.

Then the mariners were afraid, and cried every man unto his god, and cast forth the wares that were in the ship into the sea, to lighten it of them. But Jonah was gone down into the sides of the ship; and he lay, and was fast asleep.

So the shipmaster came to him, and said unto him, What meanest thou, O sleeper? arise, call upon thy God, if so be that God will think upon us, that we perish not.

And they said every one to his fellow, Come, and let us cast lots, that we may know for whose cause this evil is upon us. So they cast lots, and the lot fell upon Jonah.

Then said they unto him, Tell us, we pray thee, for whose cause this evil is upon us; What is thine occupation? and whence comest thou? what is thy country? and of what people art thou?

And he said unto them, I am an Hebrew; and I fear the Lord, the God of heaven, which hath made the sea and the dry land.

Then were the men exceedingly afraid, and said unto him. Why hast thou done this? For the men knew that he fled from the presence of the Lord, because he had told them.

Then said they unto him, What shall we do unto thee, that the sea may be calm unto us? for the sea wrought, and was tempestuous.

And he said unto them, Take me up, and cast me forth into the sea; so shall the sea be calm unto you: for I know that for my sake this great tempest is upon you.

Nevertheless the men rowed hard to bring it to the land; but they could not: for the sea wrought, and was tempestuous against them.

Wherefore they cried unto the Lord, and said, We beseech thee, O Lord, we beseech thee, let us not perish for this man's life, and lay not upon us innocent blood: for thou, O Lord, hast done as it pleased thee.

So they took up Jonah, and cast him forth into the sea: and the sea ceased from her raging.

Then the men feared the Lord exceedingly, and offered a sacrifice unto the Lord, and made vows.

Now the Lord had prepared a great fish to swallow up Jonah. And Jonah was in the belly of the fish three days and three nights.

Jonah 2

Then Jonah prayed unto the Lord his God out of the fish's belly,

And said, I cried by reason of mine affliction unto the Lord, and he heard me; out of the belly of hell cried I, and thou heardest my voice.

For thou hadst cast me into the deep, in the midst of the seas; and the floods compassed me about: all thy billows and thy waves passed over me.

Then I said, I am cast out of thy sight; yet I will look again toward thy holy temple.

The waters compassed me about, even to the soul: the depth closed me round about, the weeds were wrapped about my head.

I went down to the bottoms of the mountains; the earth with her bars was about me for ever: yet hast thou brought up my life from corruption, O Lord my God.

When my soul fainted within me I remembered the Lord: and my prayer came in unto thee, into thine holy temple.

They that observe lying vanities forsake their own mercy. But I will sacrifice unto thee with the voice of thanksgiving; I will pay that that I have vowed. Salvation is of the Lord.

And the Lord spake unto the fish, and it vomited out Jonah upon the dry land.

ABOVE: Border illustration depicting Jonah cast from the ship into the mouth of a whale. Prayer Book of Cardinal Albrecht of Brandenburg, Simon Bening, *c.* 1525–1530.
© Courtesy of The J. Paul Getty Museum, Los Angeles, Ms. Ludwig IX 19, fol. 329, 83.ML.115.329

The Book of Jonah by Unknown 49

BEOWULF

BY UNKNOWN
(C. AD 1000)

Beowulf is an epic poem in Old English, set in 5th- and 6th-century Scandinavia, about the eponymous hero of Geats (a North Germanic tribe from modern-day Sweden), known for his prowess in battle and swimming abilities. Beowulf is summoned by Hrothgar, king of the Danes, whose mead-hall is under attack by the monster Grendel. Beowulf kills Grendel in unarmed combat, then seeks out his mother, who lives in a cavern at the bottom of a lake. Beowulf swims to the bottom of the lake while withstanding attacks from monsters and serpents and kills Grendel's mother with a magic sword. Because of her lacustrine habitat, Grendel's mother is often depicted as an aquatic monster.

The poem is contained in a manuscript called the Nowell Codex, which is dated to shortly after AD 1000. It is likely that the poem was orally transmitted before it was written down, hence the presence of pagan elements, which would have existed in the original, alongside the Christian elements, which would have been introduced by the monk or monks who transcribed the story.

There are hundreds of verse translations of *Beowulf*, some of which are word for word, some archaising, some aiming to imitate the alliterative verse. The version of *Beowulf* that follows is a highly abridged prose translation from *The Red Book of Animal Stories* selected and edited by the Scottish folklorist Andrew Lang (1899). The third part of the original poem is presented in Lang's collection as a separate story titled 'The Story of Beowulf and the Fire Drake'.

LEFT: 'Grendell's mother drags Beowulf to the bottom of the lake', illustration by Henry Justice Ford. Originally published in *The Red Book of Animal Stories*, edited by Andrew Lang, 1899.
© Courtesy of Wikimedia Commons

The Story of Beowulf, Grendel, and Grendel's Mother

Long, long ago, perhaps nearly a thousand years before the adventures of the Knight of Rhodes of whom you have just heard, there lived a King of Denmark called Hrothgar. That is a curious name, you may think; but you can recognise it in our own word 'Roger,' which, of course, is common enough. This King lived in a palace, called Heorot, a princely abode, beyond what the sons of men had ever heard of; he had a beautiful wife called Waltheow, and gold, silver, and riches in abundance were his; moreover as his knights, earls, and retainers were all devotedly fond of him, he seemed to have everything in the world which could make him happy. In those days, when feasts were being held in the great halls, it was customary for one who was called a 'skald'—that is, a poet or minstrel—to sing or recite poems before the assembled company. On one of these occasions the 'skald' made poems about all sorts of evil things, wicked spirits, demons who abode in darkness, giants, ghosts, and sin and wickedness generally. It was, perhaps, not quite the sort of song to make merry the hearts of the feasters, and, in fact, it had the opposite effect, for they broke up ill at ease, as if some deadly peril were in store; nor were their presentiments without reason. That night there came to the Palace a monstrous and superhuman being named Grendel, who was the very incarnation of all cruelty and malice. He was a creature of enormous strength and size; for we read later in the story that it required four men to carry his head when he was dead. He lived an evil life, and wandered about, a lone dweller in moors, marshes, and in the wilderness. Savage and fierce as he was, nothing exasperated him more than that the King and his people should be so happy; the sound of joy and revelry within the Palace was to him as gall and wormwood. That very night, therefore, when the skald recited his ominous poem, Grendel loft his fens and marshes, and came silently to the Palace, where he found the Danes all asleep. Thirty of them he killed, devouring fifteen in the hall itself, and carrying off the rest to the marshes. Despair there was and lamentation in the morning when the other Danes arose from sleep; but none knew, or could even suggest, what was best to be done. For twelve years were the people grievously afflicted by the cruel Grendel, 'the grim stranger, the mighty haunter of the marshes, the dwelling of this monster race.' He persecuted them right sorely, nor would he have peace with any man of the Danish power. A dark, deadly shadow, he attacked alike tried warriors and youths, he ambushed and plotted, roaming the night long over the misty moors, contriving evil in his heart continually.

Matters, then, were at this pass, when a neighbouring King called Hygelac heard of the Danes' misfortunes. Hygelac reigned over the Jutes in Gotland, and he had a nephew called Beowulf, who, in common with the King and the rest of the people, was distressed to think of Hrothgar's troubles. So Beowulf made him ready a good sea-boat, took fourteen of the bravest

men-at-arms as his comrades, and set sail to help Hrothgar and the Danes. When the Danish King was told of Beowulf's arrival, he was, as you may well suppose, only too delighted, and hailed him as a heaven-sent champion, for he already knew all about him, how valiant he was, and how strong; 'for,' said Hrothgar to his people, 'it used to be said by seafaring men that this fearless warrior had in his grip the strength of thirty men.' When Beowulf came before Hrothgar, he told him, what the King already knew, that often before he had encountered sea-monsters, destroyed the Jotun tribe and slain night Nixes; and that hitherto all his deeds of prowess had been successful.

'I hear,' he said, 'that Grendel, from the thickness of his hide, cares not for weapons; I therefore disdain to carry sword or shield into the combat, but with hand-grips will I lay hold on the foe, and fight for life, man to man.'

Beowulf ended by asking that his 'garments of battle' might be sent back to his lord and kinsman Hygelac, if Grendel proved victorious in the fight. The King relied with steadfast faith upon his guest; there was now joy in the Palace of Heorot, and Queen Waltheow herself, golden-wreathed, came forth to greet the men in the hall; to each she gave a costly cup—to each his several share—'until it befell that she, the neck-laced Queen, gentle in manners and mind, bare the mead-cup to Beowulf,' and thanked God that she might find any to trust to for relief in her troubles. They all retired to rest; but not one of Beowulf's comrades thought that they would escape alive, or get them thence in safety to their well-loved homes.

That night from the moor, under the misty slopes, came Grendel prowling; in the gloom he came to the Palace, where the men-at-arms slept, whose duty it was to guard the battlemented hall; they slept, all save one. With his vast strength the monster burst open the door, and strode forward, his eyes blazing like fire. With a grim smile of delight he saw the sleepers, seized one of them and devoured him all but the feet and hands. Then he reached out at Beowulf, but the warrior clasped the extended hand and firmly grappled with the enemy. A battle royal ensued; the hall resounded with cries and shrieks, for the Danes were roused from their slumbers. They tried to help Beowulf with swords and other weapons, not knowing that they were of no avail against the monster. But the Jute yielded never a whit, he pressed Grendel harder and harder with that mighty hand-grip of his, and by sheer strength tore off the monster's hand, arm, and shoulder. Grendel fled; back to the lake he went, to the Nixes' mere, where the water for days afterwards was troubled and discoloured with blood.

As for Beowulf, the grateful King could hardly thank him enough. A feast was prepared, the walls of the great hall were covered with cloth of gold,

Beowulf by Unknown   53

and the hero received a war-banner, helmet, and breastplate, besides golden cups, a superb golden collar, and many other precious things. When the banquet was over they all retired to rest, as they supposed, in safety. But an avenger was at hand, Grendel's mother, a monstrous witch, ravenous, wrathful, and cruel as her son. She burst into Heorot, seized the man who was the King's favourite amongst all his nobles, and carried him off to the lake. She also took with her Grendel's blood-stained hand, which had been put up as a trophy. Beowulf was not in the Palace at the time, for another lodging had been given to him; but he was quickly summoned after this new disaster.

'Never fear,' said he, 'I promise thee she shall not escape, neither by water, nor into the earth, nor into the mountain forest, nor into the bottom of the sea, let her go where she will.' So they made ready at once to go to the lake, which was about a mile from the Palace; a gloomy water it was, overhung with trees, and how deep none had ever found out; every night, men said, a strange fire was to be seen on its surface, so none cared about going there. However, the King's horse was now saddled, and his men-at-arms were ready; Beowulf put on armour to protect his body from the enemy's grip, and a white helmet guarded his head. One of Hrothgar's men lent him a short sword that had never yet failed anyone who had used it in battle. Then the expedition started: over a steep and stony rise through narrow roads, past precipitous headlands they went, till they came to a bare rock and a cheerless wood, below which lay the water, dreary and troubled. They were maddened with rage when they saw the head of Æschere lying on the ground; he was the noble taken by Grendel's mother.

The water of the lake was bubbling with blood; many strange creatures of the serpent kind glided over the surface, and the men could also see Nixes lying on the headland slopes. Beowulf shot at one of the horrid water creatures with an arrow, wounding it only; but the King's men pursued it with poles and battle-axes, and killed it. Then Beowulf asked Hrothgar to send back all his presents to Hygelac, if it should happen that he, Beowulf, perished in the water. Hastening away, he plunged into the lake, and it was not very long before Grendel's mother found out that some man from above had invaded her dwelling. She grappled with him in her dreadful grasp, endeavouring to crush him to death, but his chain-mail protected him. Then she dragged him down to her den at the bottom; but meanwhile many strange beasts with terrible tusks pressed him hard in those depths, one of them even rent his war-shirt with its talons. Beowulf found himself in some kind of dreadful hall, where no water seemed to touch him; the light of a fire, a glittering ray, lit up the cavern. He could now clearly distinguish the mighty lake-witch, and thrust strongly at her with his war sword, which rang out shrilly on her head. But, alas! its edge would not bite; she had probably bewitched it with spells, as often happened in old days. So Beowulf threw away his sword, and came to close grips with her, trusting in his mighty strength. He seized her by the shoulder, but unluckily tripped and fell. In a moment she was upon him, and seized her broad dagger with deadly intent. Then, indeed, had it gone hard with Beowulf but for his coat of chain-mail, which protected his shoulder from the furious blow she gave. Suddenly he saw lying on the floor a magic sword; a huge weapon with finest edge, forged of old in the time of the Jotuns, or giants, whose work it was. No ordinary man could have wielded that blade, but Beowulf seized it, and smote the witch a fearful blow, almost cleaving her body in twain. A bright light shone up at once in the cavern, which the warrior now began to explore; nor had he gone far before he found Grendel lying on a couch, dead, so Beowulf cut off his head.

Meanwhile Hrothgar and the rest of the Danes had been sitting watching the water, which suddenly became thick and stained with blood; they had no hope that Beowulf survived. What, then, was their astonishment and delight to see him swimming towards them, breasting the waves with mighty strokes, and bearing the head of Grendel with him. And now a marvel befell; the sword with which Grendel's mother had been slain began slowly to melt away, just like ice; for the hag's blood was of such power that it consumed the blade, until nothing was left but the hilt, which was of gold, richly chased, and carved with strange characters called 'runes.' Beowulf swam ashore, and gave an account of his adventures; four men, as we have already said, bore Grendel's head to the Palace, where the hilt of the magic sword was closely examined. The characters graven upon it were found to be a description of the battle between the Gods and the Frost-Giants, in which the Giants were defeated and overwhelmed in a flood. There is an account of it in an Icelandic poem, called the 'Voluspa,' or the 'Song of the Prophetess,' which describes the Northern ideas of the creation of the world; and tells how evil and death came upon man, predicts the destruction of the universe, and gives an account of the future abodes of bliss and misery. Thus did Beowulf deliver the Danes from their misfortunes, after which he returned home, and on the death of his uncle, Hygelac, became King of Gotland.

ABOVE: The first folio of the heroic epic poem *Beowulf*, written primarily in the West Saxon dialect of Old English. Part of the Cotton MS Vitellius A XV manuscript c. 975–1025, currently located within the British Library. © Courtesy of Wikimedia Commons

THE NATURAL HISTORY OF NORWAY

BY ERIK PONTOPPIDAN
(1755)

The Danish bishop Erik Ludvigsen Pontoppidan (1698–1764) was born in Aarhus, Denmark, and studied theology at the University of Copenhagen. He was a prolific writer of works on theology but is now best known for *Det første Forsøg paa Norges naturlige Historie*, The Natural History of Norway, a comprehensive account of the history, geology, flora and fauna of Norway, which was published in four books from 1752 to 1754. It was translated into English in 1755.

Part II, Chapter VIII, 'Concerning certain Sea-monsters, or strange and uncommon Sea-animals', is an account of the existence of sea monsters. Pontoppidan's approach is systematic and proto-scientific: he considers, in turn, the existence of mermaids, sea serpents and kraken from a broad range of historical and contemporary sources. His description of the kraken is referenced in *Moby-Dick* (see page 91) and *Twenty Thousand Leagues Under the Sea* (see page 123).

LEFT: Illustration of a sea serpent in *The Natural History of Norway*, published by A Linde, 1755.
© Courtesy of The Biodiversity Heritage Library, 1856340

The Natural History of Norway
Part II, Chapter VIII: Concerning certain Sea-monsters, or strange and uncommon Sea-animals

SECT. XI. Of the Kraken, Krabben or Horven, the largest of all animals.

I am now come to the third and incontestably the largest Sea-monster in the world; it is called Kraken, Kraxen, or, as some world name it, Krabben, that word being applied by way of eminence to this creature. This last name seems indeed best to agree with the description of this creature, which is round, flat, and full of arms, or branches. Others call it also Horven, or Soe-horven, and some Anker-trold. Among all the foreign writers, both ancient and modern, which I have had opportunity to consult on this subject, not one of them seems to know much of this creature, or at least to have a just idea of it. What they say however of floating islands, as they apprehended them to be, (a thing improbable that they should exist in the wild tumultuous ocean) shall afterwards be spoken of, and will be found applicable without any hyperbole to this creature, when I shall have first given some account of it. This I shall do according to what has been related to me by my correspondents, and what I have otherwise collected by an industrious enquiry and examination into every particular, concerning which I could receive intelligence. All this, in comparison to the unknown nature and construction of the creature, is very short of a perfect account, deficient, and calculated to awake rather than satisfy the reader's curiosity. Bochart might therefore with reason fay, Lib. i. cap 6, with Oppian. Halieut. cap. I. *In mari multa latent*, i.e. *In the ocean many things are hidden*. Amongst the many great things which are in the ocean, and concealed from our eyes, or only presented to our view for a few minutes, is the Kraken. This creature is the largest and most surprizing of all the animal creation, and consequently well deserves such an account as the nature of the thing, according to the Creator's wise ordinance, will admit of. Such I shall give at present, and perhaps much greater light in this subject may be reserved for posterity according to the words of the son of Sirach, "Who hath seen him, that he might tell us? and who can magnify him as he is? There are yet hid greater things than these be, for we have feen but a few of his works." Ecclus. chap, xliii. ver. 31, 32.

SECT. XII. Their description, according to the testimony of many eye-witnesses.

Cur fishermen unanimously affirm, and without the least variation in their accounts, that when they row out several miles to sea, particularly in the hot Summer days, and by their situation (which they know by taking a view of certain points of land) expect to find 80 or 100 fathoms water, it often happens that they do not find above 20 or 30, and sometimes less. At these places they generally find the greatest plenty of Fish, especially Cod and Ling. Their lines they say are no sooner out than they may draw them up with the hooks all full of Fish; by this they judge that the Kraken is at the bottom. They say this creature causes those unnatural shallows mentioned above, and prevents their sounding. These the fishermen are always glad to find, looking upon them as a means of their taking abundance of Fish. There are sometimes twenty boats or more got together, and throwing out their lines at a moderate distance from each other; and the only thing they then have to observe is, whether the depth continues the same, which they know by their lines, or whether it grows shallower by their seeming to have less water. If this last be the case, they find that the Kraken is raising himself nearer the surface, and then it is not time for them to stay any longer; they immediately leave of fishing, take to their oars, and get away as fast as they can. When they have reached the usual depth of the place, and find themselves out of danger, they lie upon their oars, and in a few minutes after they see this enormous monster come up to the surface of the water; he there shows himself sufficiently, though his whole body does not appear, which in all likelihood no human eye ever beheld (excepting the young of this species, which shall afterwards be spoken of;) its back or upper part, which seems to be in appearance about an English mile and an half in circumference, (some say more, but I chuse the least for greater certainty) looks at first like a number of small islands, surrounded with something that floats and fluctuates like sea-weeds, Here and there

ABOVE: Title page of *The Natural History of Norway*, published by A Linde, 1755. © Courtesy of The Biodiversity Heritage Library, 1856340

a larger rising is observed like sand-banks, on which various kinds of small Fifties are seen continually leaping about till they role off into the water from the sides of it; at last several bright points or horns appear, which grow thicker and thicker the higher they rise above the surface of the water, and sometimes they stand up as high and as large as the mails of middle siz'd vessels.

It seems these are the creature's arms, and, it is said, if they were to lay hold of the largest man of war, they would pull it down to the bottom. After this monster has been on the surface of the water a short time, it begins slowly to sink again, and then the danger is as great as before; because the motion of his finking causes such a swell in the sea, and such an eddy or whirlpool, that it draws everything down with it, like the current of the river Male, which has been described in its proper place. As this enormous Sea-animal in all probability may be reckon'd of the Polype, or of the Star-fish kind, as shall hereafter be more fully proved, it seems that the parts which are seen riling at its pleasure, and are called arms, are properly the tentacula, or feeling instruments, called horns as well as arms. With these they move themselves, and likewise gather in their food.

Besides these, for this last purpose the great Creator has also given this creature a strong and peculiar scent, which it can emit at certain times, and by means of which it beguiles and draws other Fish to come in heaps about it. This animal has another strange property, known by the experience of a great many old fishermen. They observe, that for some months the Kraken or Krabben is continually eating, and in other months he always voids his excrements. During this evacuation the surface of the water is

ABOVE: Erik Pontoppidan by A Brynnik, 1749.
© Courtesy of Wikimedia Commons

coloured with the excrement, and appears quite thick and turbid. This muddiness is said to be so very agreeable to the smell or taste of other Fishes, or to both, that they gather together from all parts to it, and keep for that purpose directly over the Kraken: he then opens his arms, or horns, seizes and swallows his welcome guests, and converts them, after the due time, by digestion, into a bait for other Fish of the same kind. I relate what is affirmed by many; but I cannot give so certain assurances of this particular, as I can of the existence of this surprizing creature; though I do not find any thing in it absolutely contrary to nature. As we can hardly expect an opportunity to examine this enormous sea-animal alive, I am the more concerned that nobody embraced that opportunity which, according to the following account, once did, and perhaps never more may offer, of seeing it entire when dead. The reverend Mr. Friis, consistorial assessor, minister of Bodoen in Nordland, and vicar of the college for promoting christian knowledge, gave me at the latter end of last year, when he was at Bergen, this relation; which I deliver again on his credit.

In the year 1680 a Krake (perhaps a young and careless one) came into the water that runs between the rocks and cliffs in the parish of Alstahoug, though the general custom of that creature is to keep always several leagues from land, and therefore of course they must die there. It happened that its extended long arms, or antennæ, which this creature seems to use like the Snail, in turning about, caught hold of some trees standing near the water, which might easily have been torn up by the roots but beside this, as it was found afterwards, he entangled himself in some openings or clefts in the rock, and therein stuck so fast and hung so unfortunately, that he could not work himself out, but perished and putrified on the spot. The carcase, which was a long while decaying, and filled great part of that narrow channel made it almost impassable by its intolerable stench.

The Kraken has never been known to do any great harm, except they have taken away the lives of those who consequently could not bring the tidings. I have never heard but one instance mentioned, which happened a few years ago near Fridrichstad, in the diocess of Aggerhuus. They say that two fishermen accidentally, and to their great surprize, fell into such a spot on the water as has been before described, full of a thick slime, almost like a morass. They immediately strove to get out of this place, but they had not time to turn quick enough to save themselves from one of the Kraken's horns, which crushed the head of the boat so, that it was with great difficulty they saved their lives on the wreck, tho' the weather was as calm as possible; for these monsters, like the Sea-snake, never appear at other times.

RIGHT: Illustration of a sea monster, *Historiæ insectorum Libellus*, Conradi Gesneri, 1620.
© Courtesy of Internet Archive

'THE SEA-NYMPH'
BY ANN RADCLIFFE
(1794)

Ann Radcliffe (1764–1823) was one of, if not the most well-known author of the Gothic novel. Born Ann Ward, she married the journalist William Radcliffe and they lived in London, where William wrote for a leftist newspaper. Radcliffe published her first novel, *The Castles of Athlin and Dunbayne*, in 1789, soon followed by *A Sicilian Romance* (1790), *The Romance of the Forest* (1791), *The Mysteries of Udolpho* (1794) and *The Italian* (1797). Despite her immense popularity and financial success, Radcliffe then retreated from public life and didn't publish anything for her remaining 26 years.

The Mysteries of Udolpho features what are now – owing to the novel's immense success – considered central tropes of the Gothic novel. Set in 1584, it is about a young, beautiful and virtuous French ingenue, Emily St. Aubert, who is held captive in Castle Udolpho by the brooding villain, Signor Montoni. Before this, while in Venice, Emily watches, awestruck, as a procession of Venetians dressed as Neptune and his court float along a canal. She imagines herself as a sea nymph and composes a poem.

In Greek myth, 'sea-nymphs' were Nereids: spirits or embodiments of the sea and often visually depicted as human women. In the Romantic and Victorian imagination, they became synonymous with mermaids, more often depicted with tails.

LEFT: *A Midsummer Night's Dream*, illustration by Arthur Rackham, 1908.
© Courtesy of Getty Images

'The Sea-Nymph'

Down, down a thousand fathom deep,
Among the sounding seas I go;
Play round the foot of every steep
Whose cliffs above the ocean grow.
There, within their secret caves,
I hear the mighty rivers roar;
And guide their streams through Neptune's waves
To bless the green earth's inmost shore:
And bid the freshen'd waters glide,
For fern-crown'd nymphs of lake, or brook,
Through winding woods and pastures wide,
And many a wild, romantic nook.
For this the nymphs, at fall of eve,
Oft dance upon the flow'ry banks,
And sing my name, and garlands weave
To bear beneath the wave their thanks.
In coral bow'rs I love to lie,
And hear the surges roll above,
And, through the waters, view on high
The proud ships sail, and gay clouds move.
And oft at midnight's stillest hour,
When summer seas the vessel lave,
I love to prove my charmful pow'r
While floating on the moon-light wave.
And when deep sleep the crew has bound,
And the sad lover musing leans
O'er the ship's side, I breathe around
Such strains as speak no mortal means!
O'er the dim waves his searching eye
Sees but the vessel's lengthen'd shade;
Above—the moon and azure sky;

Entranc'd he hears, and half afraid!
Sometimes a single note I swell,
That, softly sweet, at distance dies;
Then wake the magic of my shell,
And choral voices round me rise!
The trembling youth, charm'd by my strain,
Calls up the crew, who, silent, bend
O'er the high deck, but list in vain;
My song is hush'd, my wonders end!
Within the mountain's woody bay,
Where the tall bark at anchor rides,
At twilight hour, with tritons gay,
I dance upon the lapsing tides.
And with my sister-nymphs I sport,
'Till the broad sun looks o'er the floods;
Then, swift we seek our crystal court,
Deep in the wave, 'mid Neptune's woods.
In cool arcades and glassy halls
We pass the sultry hours of noon,
Beyond wherever sun-beam falls,
Weaving sea-flowers in gay festoon.
The while we chant our ditties sweet
To some soft shell that warbles near;
Join'd by the murmuring currents, fleet,
That glide along our halls so clear.
There, the pale pearl and sapphire blue,
And ruby red, and em'rald green,
Dart from the domes a changing hue,
And sparry columns deck the scene.
When the dark storm scowls o'er the deep,
And long, long peals of thunder sound,
On some high cliff my watch I keep

O'er all the restless seas around:
'Till on the ridgy wave, afar,
Comes the lone vessel, labouring slow,
Spreading the white foam in the air,
With sail and topmast bending low.
Then, plunge I 'mid the ocean's roar,
My way by quiv'ring lightnings shewn,
To guide the bark to peaceful shore,
And hush the sailor's fearful groan.
And if too late I reach its side
To save it from the 'whelming surge,
I call my dolphins o'er the tide,
To bear the crew where isles emerge.
Their mournful spirits soon I cheer,
While round the desert coast I go,
With warbled songs they faintly hear,
Oft as the stormy gust sinks low.
My music leads to lofty groves,
That wild upon the sea-bank wave;
Where sweet fruits bloom, and fresh spring roves,
And closing boughs the tempest brave.
The spirits of the air obey
My potent voice they love so well;
And, on the clouds, paint visions gay,
While strains more sweet at distance swell.
And thus the lonely hours I cheat,
Soothing the ship-wreck'd sailor's heart,
'Till from the waves the storms retreat,
And o'er the east the day-beams dart.
Neptune for this oft binds me fast
To rocks below, with coral chain,
'Till all the tempest's over-past,

ABOVE: 'All alone, those rocks amid – one night I very nearly did!'
Illustration by Harry Clarke, *The Year's at the Spring*, 1920.
© Courtesy of Wikimedia Commons

And drowning seamen cry in vain.
Whoe'er ye are that love my lay,
Come, when red sun-set tints the wave
To the still sands, where fairies play;
There, in cool seas, I love to lave.

'The Sea-Nymph' by Ann Radcliffe   65

THE RIME OF THE ANCIENT MARINER

BY SAMUEL TAYLOR COLERIDGE
(1798)

Samuel Taylor Coleridge (1772–1834), along with William Wordsworth, is considered a founder of English Romanticism. His longest poem, a ballad titled *The Rime of the Ancient Mariner*, was written in 1797–98 and published in 1798 in Coleridge and Wordsworth's collection *Lyrical Ballads*. The introductory 'Advertisement' is a manifesto of sorts of Romantic poetry.

The 'Rime', told in language that was archaic even for the time, is the tale of a shipwrecked sailor who now wanders the world, compelled to tell his story. The presence of sea monsters in the 'Rime' is understated, yet it is the mariner's changing attitude towards them that frees him of a curse invoked by wrathful spirits when he shoots an albatross. He is first disgusted by the 'slimy things' in the deep. Later, after the rest of the crew's souls are claimed by Death, the mariner blesses the 'rich attire' of the 'happy living things' he previously abhorred.

Coleridge often revised his texts for subsequent publications. This version is from 1834, the year of his death. The poem has been illustrated by many artists, most notably the French artist Gustave Doré (1832–1883), who produced 39 engraved plates and three vignettes for Dover's 1876 publication. Doré's illustrations of the tumultuous ocean and tortured mariner are regarded for their depiction of sublime awe and terror.

LEFT: 'The ship drove fast, loud roared the blast, And southward aye we fled.' Illustration by Gustave Doré, 1876.
© Courtesy of iStock

The Rime of the Ancient Mariner

Argument

How a Ship having passed the Line was driven by storms to the cold Country towards the South Pole; and how from thence she made her course to the tropical Latitude of the Great Pacific Ocean; and of the strange things that befell; and in what manner the Ancyent Marinere came back to his own Country.

PART I

It is an ancient Mariner,
And he stoppeth one of three.
'By thy long grey beard and glittering eye,
Now wherefore stopp'st thou me?

The Bridegroom's doors are opened wide,
And I am next of kin;
The guests are met, the feast is set:
May'st hear the merry din.'

He holds him with his skinny hand,
'There was a ship,' quoth he.
'Hold off! unhand me, grey-beard loon!'
Eftsoons his hand dropt he.

He holds him with his glittering eye—
The Wedding-Guest stood still,
And listens like a three years' child:
The Mariner hath his will.

The Wedding-Guest sat on a stone:
He cannot choose but hear;
And thus spake on that ancient man,
The bright-eyed Mariner.

'The ship was cheered, the harbour cleared,
Merrily did we drop
Below the kirk, below the hill,
Below the lighthouse top.

The Sun came up upon the left,
Out of the sea came he!
And he shone bright, and on the right
Went down into the sea.

Higher and higher every day,
Till over the mast at noon—'
The Wedding-Guest here beat his breast,
For he heard the loud bassoon.

The bride hath paced into the hall,
Red as a rose is she;
Nodding their heads before her goes
The merry minstrelsy.

The Wedding-Guest he beat his breast,
Yet he cannot choose but hear;
And thus spake on that ancient man,
The bright-eyed Mariner.

And now the STORM-BLAST came, and he
Was tyrannous and strong:

He struck with his o'ertaking wings,
And chased us south along.

With sloping masts and dipping prow,
As who pursued with yell and blow
Still treads the shadow of his foe,
And forward bends his head,
The ship drove fast, loud roared the blast,
And southward aye we fled.

And now there came both mist and snow,
And it grew wondrous cold:
And ice, mast-high, came floating by,
As green as emerald.

And through the drifts the snowy clifts
Did send a dismal sheen:
Nor shapes of men nor beasts we ken—
The ice was all between.

The ice was here, the ice was there,
The ice was all around:
It cracked and growled, and roared and howled,
Like noises in a swound!

At length did cross an Albatross,
Thorough the fog it came;
As if it had been a Christian soul,
We hailed it in God's name.

It ate the food it ne'er had eat,
And round and round it flew.
The ice did split with a thunder-fit;
The helmsman steered us through!

And a good south wind sprung up behind;
The Albatross did follow,

And every day, for food or play,
Came to the mariner's hollo!

In mist or cloud, on mast or shroud,
It perched for vespers nine;
Whiles all the night, through fog-smoke white,
Glimmered the white Moon-shine.'

'God save thee, ancient Mariner!
From the fiends, that plague thee thus!—
Why look'st thou so?'—With my cross-bow
I shot the ALBATROSS.

ABOVE: 'And now there came both mist and snow, And it grew wondrous cold.' Illustration by Gustave Doré, 1876.
© Courtesy of iStock

PART II

The Sun now rose upon the right:
Out of the sea came he,
Still hid in mist, and on the left
Went down into the sea.

And the good south wind still blew behind,
But no sweet bird did follow,
Nor any day for food or play
Came to the mariner's hollo!

And I had done a hellish thing,
And it would work 'em woe:
For all averred, I had killed the bird
That made the breeze to blow.
Ah wretch! said they, the bird to slay,
That made the breeze to blow!

Nor dim nor red, like God's own head,
The glorious Sun uprist:
Then all averred, I had killed the bird
That brought the fog and mist.
'Twas right, said they, such birds to slay,
That bring the fog and mist.

The fair breeze blew, the white foam flew,
The furrow followed free;
We were the first that ever burst
Into that silent sea.

Down dropt the breeze, the sails dropt down,
'Twas sad as sad could be;
And we did speak only to break
The silence of the sea!

All in a hot and copper sky,
The bloody Sun, at noon,
Right up above the mast did stand,
No bigger than the Moon.

Day after day, day after day,
We stuck, nor breath nor motion;
As idle as a painted ship
Upon a painted ocean.

Water, water, every where,
And all the boards did shrink;
Water, water, every where,
Nor any drop to drink.

The very deep did rot: O Christ!
That ever this should be!
Yea, slimy things did crawl with legs
Upon the slimy sea.

About, about, in reel and rout
The death-fires danced at night;
The water, like a witch's oils,
Burnt green, and blue and white.

And some in dreams assurèd were
Of the Spirit that plagued us so;
Nine fathom deep he had followed us
From the land of mist and snow.

And every tongue, through utter drought,
Was withered at the root;
We could not speak, no more than if
We had been choked with soot.

Ah! well a-day! what evil looks
Had I from old and young!

Instead of the cross, the Albatross
About my neck was hung.

PART III

There passed a weary time. Each throat
Was parched, and glazed each eye.
A weary time! a weary time!
How glazed each weary eye,

When looking westward, I beheld
A something in the sky.

At first it seemed a little speck,
And then it seemed a mist;
It moved and moved, and took at last
A certain shape, I wist.

A speck, a mist, a shape, I wist!
And still it neared and neared:
As if it dodged a water-sprite,
It plunged and tacked and veered.

With throats unslaked, with black lips baked,
We could nor laugh nor wail;
Through utter drought all dumb we stood!
I bit my arm, I sucked the blood,
And cried, A sail! a sail!

With throats unslaked, with black lips baked,
Agape they heard me call:
Gramercy! they for joy did grin,
And all at once their breath drew in.
As they were drinking all.

See! see! (I cried) she tacks no more!
Hither to work us weal;

Without a breeze, without a tide,
She steadies with upright keel!

The western wave was all a-flame.
The day was well nigh done!
Almost upon the western wave
Rested the broad bright Sun;
When that strange shape drove suddenly
Betwixt us and the Sun.

And straight the Sun was flecked with bars,
(Heaven's Mother send us grace!)
As if through a dungeon-grate he peered
With broad and burning face.

Alas! (thought I, and my heart beat loud)
How fast she nears and nears!
Are those *her* sails that glance in the Sun,
Like restless gossameres?

Are those her *ribs* through which the Sun
Did peer, as through a grate?
And is that Woman all her crew?
Is that a DEATH? and are there two?
Is DEATH that woman's mate?

Her lips were red, *her* looks were free,
Her locks were yellow as gold:
Her skin was as white as leprosy,
The Night-mare LIFE-IN-DEATH was she,
Who thicks man's blood with cold.

The naked hulk alongside came,
And the twain were casting dice;
'The game is done! I've won! I've won!'
Quoth she, and whistles thrice.

The Sun's rim dips; the stars rush out;
At one stride comes the dark;
With far-heard whisper, o'er the sea,
Off shot the spectre-bark.

We listened and looked sideways up!
Fear at my heart, as at a cup,
My life-blood seemed to sip!
The stars were dim, and thick the night,
The steersman's face by his lamp gleamed white;
From the sails the dew did drip—
Till clomb above the eastern bar
The hornèd Moon, with one bright star
Within the nether tip.

One after one, by the star-dogged Moon,
Too quick for groan or sigh,
Each turned his face with a ghastly pang,
And cursed me with his eye.

Four times fifty living men,
(And I heard nor sigh nor groan)
With heavy thump, a lifeless lump,
They dropped down one by one.

The souls did from their bodies fly,—
They fled to bliss or woe!
And every soul, it passed me by,
Like the whizz of my cross-bow!

PART IV

'I fear thee, ancient Mariner!
I fear thy skinny hand!
And thou art long, and lank, and brown,
As is the ribbed sea-sand.

I fear thee and thy glittering eye,
And thy skinny hand, so brown.'—
Fear not, fear not, thou Wedding-Guest!
This body dropt not down.

Alone, alone, all, all alone,
Alone on a wide wide sea!
And never a saint took pity on
My soul in agony.

The many men, so beautiful!
And they all dead did lie:
And a thousand thousand slimy things
Lived on; and so did I.

I looked upon the rotting sea,
And drew my eyes away;
I looked upon the rotting deck,
And there the dead men lay.

ABOVE: 'Beyond the shadow of the ship, I watched the water-snakes.' Illustration by Gustave Doré, 1876.
© Courtesy of iStock

I looked to heaven, and tried to pray;
But or ever a prayer had gusht,
A wicked whisper came, and made
My heart as dry as dust.

I closed my lids, and kept them close,
And the balls like pulses beat;
For the sky and the sea, and the sea and the sky
Lay dead like a load on my weary eye,
And the dead were at my feet.

The cold sweat melted from their limbs,
Nor rot nor reek did they:
The look with which they looked on me
Had never passed away.

An orphan's curse would drag to hell
A spirit from on high;
But oh! more horrible than that
Is the curse in a dead man's eye!
Seven days, seven nights, I saw that curse,
And yet I could not die.

The moving Moon went up the sky,
And no where did abide:
Softly she was going up,
And a star or two beside—

Her beams bemocked the sultry main,
Like April hoar-frost spread;
But where the ship's huge shadow lay,
The charmèd water burnt alway
A still and awful red.

Beyond the shadow of the ship,
I watched the water-snakes:
They moved in tracks of shining white,

ABOVE: 'The game is done! I've won, I've won!' Illustration by Gustave Doré, 1876.
© Courtesy of iStock

And when they reared, the elfish light
Fell off in hoary flakes.

Within the shadow of the ship
I watched their rich attire:
Blue, glossy green, and velvet black,
They coiled and swam; and every track
Was a flash of golden fire.

O happy living things! no tongue
Their beauty might declare:
A spring of love gushed from my heart,
And I blessed them unaware:
Sure my kind saint took pity on me,
And I blessed them unaware.

The self-same moment I could pray;
And from my neck so free
The Albatross fell off, and sank
Like lead into the sea.

PART V

Oh sleep! it is a gentle thing,
Beloved from pole to pole!
To Mary Queen the praise be given!
She sent the gentle sleep from Heaven,
That slid into my soul.

The silly buckets on the deck,
That had so long remained,
I dreamt that they were filled with dew;
And when I awoke, it rained.

My lips were wet, my throat was cold,
My garments all were dank;
Sure I had drunken in my dreams,
And still my body drank.

I moved, and could not feel my limbs:
I was so light—almost
I thought that I had died in sleep,
And was a blessed ghost.

And soon I heard a roaring wind:
It did not come anear;
But with its sound it shook the sails,
That were so thin and sere.

The upper air burst into life!
And a hundred fire-flags sheen,
To and fro they were hurried about!
And to and fro, and in and out,
The wan stars danced between.

And the coming wind did roar more loud,
And the sails did sigh like sedge,
And the rain poured down from one black cloud;
The Moon was at its edge.

The thick black cloud was cleft, and still
The Moon was at its side:
Like waters shot from some high crag,
The lightning fell with never a jag,
A river steep and wide.

The loud wind never reached the ship,
Yet now the ship moved on!
Beneath the lightning and the Moon
The dead men gave a groan.

They groaned, they stirred, they all uprose,
Nor spake, nor moved their eyes;
It had been strange, even in a dream,
To have seen those dead men rise.

The helmsman steered, the ship moved on;
Yet never a breeze up-blew;
The mariners all 'gan work the ropes,
Where they were wont to do;
They raised their limbs like lifeless tools—
We were a ghastly crew.

The body of my brother's son
Stood by me, knee to knee:
The body and I pulled at one rope,
But he said nought to me.

'I fear thee, ancient Mariner!'
Be calm, thou Wedding-Guest!
'Twas not those souls that fled in pain,
Which to their corses came again,
But a troop of spirits blest:

For when it dawned—they dropped their arms,
And clustered round the mast;
Sweet sounds rose slowly through their mouths,
And from their bodies passed.

Around, around, flew each sweet sound,
Then darted to the Sun;
Slowly the sounds came back again,
Now mixed, now one by one.

Sometimes a-dropping from the sky
I heard the sky-lark sing;
Sometimes all little birds that are,
How they seemed to fill the sea and air
With their sweet jargoning!

And now 'twas like all instruments,
Now like a lonely flute;
And now it is an angel's song,
That makes the heavens be mute.

It ceased; yet still the sails made on
A pleasant noise till noon,
A noise like of a hidden brook
In the leafy month of June,
That to the sleeping woods all night
Singeth a quiet tune.

Till noon we quietly sailed on,
Yet never a breeze did breathe:
Slowly and smoothly went the ship,
Moved onward from beneath.

Under the keel nine fathom deep,
From the land of mist and snow,
The spirit slid: and it was he
That made the ship to go.

The sails at noon left off their tune,
And the ship stood still also.

The Sun, right up above the mast,
Had fixed her to the ocean:
But in a minute she 'gan stir,
With a short uneasy motion—
Backwards and forwards half her length
With a short uneasy motion.

Then like a pawing horse let go,
She made a sudden bound:
It flung the blood into my head,
And I fell down in a swound.

How long in that same fit I lay,
I have not to declare;
But ere my living life returned,
I heard and in my soul discerned
Two voices in the air.

'Is it he?' quoth one, 'Is this the man?
By him who died on cross,
With his cruel bow he laid full low
The harmless Albatross.

The spirit who bideth by himself
In the land of mist and snow,
He loved the bird that loved the man
Who shot him with his bow.'

The other was a softer voice,
As soft as honey-dew:
Quoth he, 'The man hath penance done,
And penance more will do.'

The Rime of the Ancient Mariner by Samuel Taylor Coleridge

PART VI

First Voice
'But tell me, tell me! speak again,
Thy soft response renewing—
What makes that ship drive on so fast?
What is the ocean doing?'

Second Voice
Still as a slave before his lord,
The ocean hath no blast;
His great bright eye most silently
Up to the Moon is cast—

If he may know which way to go;
For she guides him smooth or grim.
See, brother, see! how graciously
She looketh down on him.'

First Voice
'But why drives on that ship so fast,
Without or wave or wind?'

Second Voice
'The air is cut away before,
And closes from behind.

Fly, brother, fly! more high, more high!
Or we shall be belated:
For slow and slow that ship will go,
When the Mariner's trance is abated.'

I woke, and we were sailing on
As in a gentle weather:
'Twas night, calm night, the moon was high;
The dead men stood together.

All stood together on the deck,
For a charnel-dungeon fitter:
All fixed on me their stony eyes,
That in the Moon did glitter.

The pang, the curse, with which they died,
Had never passed away:
I could not draw my eyes from theirs,
Nor turn them up to pray.

And now this spell was snapt: once more
I viewed the ocean green,
And looked far forth, yet little saw
Of what had else been seen—

Like one, that on a lonesome road
Doth walk in fear and dread,
And having once turned round walks on,
And turns no more his head;
Because he knows, a frightful fiend
Doth close behind him tread.

But soon there breathed a wind on me,
Nor sound nor motion made:
Its path was not upon the sea,
In ripple or in shade.

It raised my hair, it fanned my cheek
Like a meadow-gale of spring—
It mingled strangely with my fears,
Yet it felt like a welcoming.

Swiftly, swiftly flew the ship,
Yet she sailed softly too:
Sweetly, sweetly blew the breeze—
On me alone it blew.

Oh! dream of joy! is this indeed
The light-house top I see?
Is this the hill? is this the kirk?
Is this mine own countree?

We drifted o'er the harbour-bar,
And I with sobs did pray—
O let me be awake, my God!
Or let me sleep alway.

The harbour-bay was clear as glass,
So smoothly it was strewn!
And on the bay the moonlight lay,
And the shadow of the Moon.

The rock shone bright, the kirk no less,
That stands above the rock:
The moonlight steeped in silentness
The steady weathercock.

And the bay was white with silent light,
Till rising from the same,
Full many shapes, that shadows were,
In crimson colours came.

A little distance from the prow
Those crimson shadows were:
I turned my eyes upon the deck—
Oh, Christ! what saw I there!

Each corse lay flat, lifeless and flat,
And, by the holy rood!
A man all light, a seraph-man,
On every corse there stood.

This seraph-band, each waved his hand:
It was a heavenly sight!

ABOVE: Samuel Coleridge, age 42, 1842.
© Courtesy of Wikimedia Commons

They stood as signals to the land,
Each one a lovely light;

This seraph-band, each waved his hand,
No voice did they impart—
No voice; but oh! the silence sank
Like music on my heart.

But soon I heard the dash of oars,
I heard the Pilot's cheer;
My head was turned perforce away
And I saw a boat appear.

The Pilot and the Pilot's boy,
I heard them coming fast:
Dear Lord in Heaven! it was a joy
The dead men could not blast.

The Rime of the Ancient Mariner by Samuel Taylor Coleridge

I saw a third—I heard his voice:
It is the Hermit good!
He singeth loud his godly hymns
That he makes in the wood.
He'll shrieve my soul, he'll wash away
The Albatross's blood.

PART VII

This Hermit good lives in that wood
Which slopes down to the sea.
How loudly his sweet voice he rears!
He loves to talk with marineres
That come from a far countree.

He kneels at morn, and noon, and eve—
He hath a cushion plump:
It is the moss that wholly hides
The rotted old oak-stump.

The skiff-boat neared: I heard them talk,
'Why, this is strange, I trow!
Where are those lights so many and fair,
That signal made but now?'

'Strange, by my faith!' the Hermit said—
'And they answered not our cheer!
The planks looked warped! and see those sails,
How thin they are and sere!
I never saw aught like to them,
Unless perchance it were

Brown skeletons of leaves that lag
My forest-brook along;
When the ivy-tod is heavy with snow,
And the owlet whoops to the wolf below,
That eats the she-wolf's young.'

'Dear Lord! it hath a fiendish look—
(The Pilot made reply)
I am a-feared'—'Push on, push on!'
Said the Hermit cheerily.

The boat came closer to the ship,
But I nor spake nor stirred;
The boat came close beneath the ship,
And straight a sound was heard.

Under the water it rumbled on,
Still louder and more dread:
It reached the ship, it split the bay;
The ship went down like lead.

Stunned by that loud and dreadful sound,
Which sky and ocean smote,
Like one that hath been seven days drowned
My body lay afloat;
But swift as dreams, myself I found
Within the Pilot's boat.

Upon the whirl, where sank the ship,
The boat spun round and round;
And all was still, save that the hill
Was telling of the sound.

I moved my lips—the Pilot shrieked
And fell down in a fit;
The holy Hermit raised his eyes,
And prayed where he did sit.

I took the oars: the Pilot's boy,
Who now doth crazy go,
Laughed loud and long, and all the while
His eyes went to and fro.
'Ha! ha!' quoth he, 'full plain I see,

The Devil knows how to row.'

And now, all in my own countree,
I stood on the firm land!
The Hermit stepped forth from the boat,
And scarcely he could stand.

'O shrieve me, shrieve me, holy man!'
The Hermit crossed his brow.
'Say quick,' quoth he, 'I bid thee say—
What manner of man art thou?'

Forthwith this frame of mine was wrenched
With a woful agony,
Which forced me to begin my tale;
And then it left me free.

Since then, at an uncertain hour,
That agony returns:
And till my ghastly tale is told,
This heart within me burns.

I pass, like night, from land to land;
I have strange power of speech;
That moment that his face I see,
I know the man that must hear me:
To him my tale I teach.

What loud uproar bursts from that door!
The wedding-guests are there:
But in the garden-bower the bride
And bride-maids singing are:
And hark the little vesper bell,
Which biddeth me to prayer!

O Wedding-Guest! this soul hath been
Alone on a wide wide sea:
So lonely 'twas, that God himself
Scarce seemèd there to be.

O sweeter than the marriage-feast,
'Tis sweeter far to me,
To walk together to the kirk
With a goodly company!—

To walk together to the kirk,
And all together pray,
While each to his great Father bends,
Old men, and babes, and loving friends
And youths and maidens gay!

Farewell, farewell! but this I tell
To thee, thou Wedding-Guest!
He prayeth well, who loveth well
Both man and bird and beast.

He prayeth best, who loveth best
All things both great and small;
For the dear God who loveth us,
He made and loveth all.

The Mariner, whose eye is bright,
Whose beard with age is hoar,
Is gone: and now the Wedding-Guest
Turned from the bridegroom's door.

He went like one that hath been stunned,
And is of sense forlorn:
A sadder and a wiser man,
He rose the morrow morn.

'THE KRAKEN', 'THE MERMAN' AND 'THE MERMAID'

BY ALFRED TENNYSON
(1830)

Alfred Tennyson, 1st Baron Tennyson (1809–1892) is often considered the most prominent poet of the Victorian era. In 1850 he succeeded William Wordsworth as Poet Laureate, a role he held until his death.

Tennyson was raised as part of a large family in Somersby, in Lincolnshire, England, where his father was a clergyman. He attended King Edward VI Grammar School then Trinity College, Cambridge, publishing his first solo collection of poems – titled *Poems, Chiefly Lyrical*, in 1830. This collection included 'The Kraken', 'The Merman' and 'The Mermaid', among other lyrics, sonnets and ballads on love, mourning and mortality. 'The Kraken' is considered a sonnet but follows an unusual rhyme scheme and has an extra 15th, hexametric line. The poem, like the titular kraken, threatens to break free from its confines.

Tennyson scholars identify Pontoppidan's description of the kraken as a source of inspiration (see *The Natural History of Norway*, page 55). Likewise, scholars of HP Lovecraft identify Tennyson's poem as a source of inspiration for Lovecraft's 'The Call of Cthulhu'.

LEFT: *Le Poulpe Colossal*, Pierre Dénys de Monfort, 1801.
© Courtesy of The Biodiversity Heritage Library

'The Kraken'

Below the thunders of the upper deep;
Far, far beneath in the abysmal sea,
His ancient, dreamless, uninvaded sleep
The Kraken sleepeth: faintest sunlights flee
About his shadowy sides: above him swell
Huge sponges of millennial growth and height;
And far away into the sickly light,
From many a wondrous grot and secret cell
Unnumber'd and enormous polypi
Winnow with giant arms the slumbering green.
There hath he lain for ages and will lie
Battening upon huge seaworms in his sleep,
Until the latter fire shall heat the deep;
Then once by man and angels to be seen,
In roaring he shall rise and on the surface die.

'The Merman'

Who would be
A merman bold,
Sitting alone,
Singing alone
Under the sea,
With a crown of gold,
On a throne?
I would be a merman bold,
I would sit and sing the whole of the day;
I would fill the sea-halls with a voice of power;
But at night I would roam abroad and play
With the mermaids in and out of the rocks,
Dressing their hair with the white sea-flower;
And holding them back by their flowing locks

I would kiss them often under the sea,
And kiss them again till they kiss'd me
Laughingly, laughingly;
And then we would wander away, away,
To the pale-green sea-groves straight and high,
Chasing each other merrily.

There would be neither moon nor star;
But the wave would make music above us afar —
Low thunder and light in the magic night —
Neither moon nor star.
We would call aloud in the dreamy dells,
Call to each other and whoop and cry
All night, merrily, merrily;
They would pelt me with starry spangles and shells,
Laughing and clapping their hands between,
All night, merrily, merrily,

But I would throw to them back in mine
Turkis and agate and almondine;
Then leaping out upon them unseen
I would kiss them often under the sea,
And kiss them again till they kiss'd me
Laughingly, laughingly.
Oh! what a happy life were mine
Under the hollow-hung ocean green!
Soft are the moss-beds under the sea;
We would live merrily, merrily.

'The Kraken', 'The Merman' and 'The Mermaid' by Alfred Tennyson

'The Mermaid'

Who would be
A mermaid fair,
Singing alone,
Combing her hair
Under the sea,
In a golden curl
With a comb of pearl,
On a throne?

I would be a mermaid fair;
I would sing to myself the whole of the day;
With a comb of pearl I would comb my hair;
And still as I comb'd I would sing and say,
Who is it loves me? who loves not me?
I would comb my hair till my ringlets would fall
Low adown, low adown,
From under my starry sea-bud crown
Low adown and around,
And I should look like a fountain of gold
Springing alone
With a shrill inner sound,
Over the throne
In the midst of the hall;
Till that great sea-snake under the sea
From his coiled sleeps in the central deeps
Would slowly trail himself sevenfold
Round the hall where I sate, and look in at the gate
With his large calm eyes for the love of me.
And all the mermen under the sea
Would feel their immortality
Die in their hearts for the love of me.
But at night I would wander away, away,
I would fling on each side my low-flowing locks,
And lightly vault from the throne and play
With the mermen in and out of the rocks;
We would run to and fro, and hide and seek,
On the broad sea-wolds in the crimson shells,
Whose silvery spikes are nighest the sea.
But if any came near I would call, and shriek,
And adown the steep like a wave I would leap
From the diamond-ledges that jut from the dells;
For I would not be kiss'd by all who would list,
Of the bold merry mermen under the sea;
They would sue me, and woo me, and flatter me,
In the purple twilights under the sea;
But the king of them all would carry me,
Woo me, and win me, and marry me,
In the branching jaspers under the sea;
Then all the dry pied things that be
In the hueless mosses under the sea
Would curl round my silver feet silently,
All looking up for the love of me.
And if I should carol aloud, from aloft
All things that are forked, and horned, and soft
Would lean out from the hollow sphere
    of the sea,
All looking down for the love of me.

84 'The Kraken', 'The Merman' and 'The Mermaid' by Alfred Tennyson

RIGHT: Caricature of Tennyson from *Vanity Fair*, 22 July 1871.
© Courtesy of iStock

'THE CITY IN THE SEA'
BY EDGAR ALLAN POE
(1831)

Edgar Allan Poe (1809–1849) was born in Boston, Massachusetts. In adulthood he lived in Baltimore then Richmond with his aunt and young cousin, Virginia. In 1827 he published his first collection of poems, *Tamerlane and Other Poems*, followed by a second collection, Al Aaraaf, *Tamerlane and Minor Poems*, in 1829. He then began to write short stories for magazines.

In 1836, at the age of 26, Poe married Virginia, who was 13. She died from tuberculosis 11 years later. Poe's own death is shrouded in mystery; various theories have been put forward, including murder, suicide, cooping (involuntary participation in election fraud), rabies, syphilis and alcoholism.

Certain Gothic motifs recur across Poe's short stories and poems, such as doppelgangers, premature burial, addiction and madness. Only a few of his works explicitly feature the sea, notably the short story 'A Descent into the Maelström' (1841) and his only full-length novel, *The Narrative of Arthur Gordon Pym of Nantucket* (1838). 'The City in the Sea' (1845) is a sweeping depiction of an undersea city ruled over by the personification of death. An earlier version of the poem, titled 'The Doomed City', appeared in Poe's 1831 collection *Poems*. It was then published as 'The City of Sin' in the *Southern Literary Messenger* in 1836, and as 'The City in the Sea' in the *American Review* and *Broadway Journal* in 1845.

LEFT: 'The City in the Sea' from *The Poems of Edgar Allan Poe*, illustrated by William Heath Robinson, 1900.
© Courtesy of Internet Archive

'The City in the Sea'

Lo! Death has reared himself a throne
In a strange city lying alone
Far down within the dim West,
Where the good and the bad and the worst and the best
Have gone to their eternal rest.
There shrines and palaces and towers
(Time-eaten towers that tremble not!)
Resemble nothing that is ours.
Around, by lifting winds forgot,
Resignedly beneath the sky
The melancholy waters lie.
No rays from the holy heaven come down
On the long night-time of that town;
But light from out the lurid sea
Streams up the turrets silently —
Gleams up the pinnacles far and free —
Up domes — up spires — up kingly halls —
Up fanes — up Babylon-like walls —
Up shadowy long-forgotten bowers
Of sculptured ivy and stone flowers —
Up many and many a marvellous shrine
Whose wreathed friezes intertwine
The viol, the violet, and the vine.
Resignedly beneath the sky
The melancholy waters lie.
So blend the turrets and shadows there
That all seem pendulous in air,
While from a proud tower in the town
Death looks gigantically down.
There open fanes and gaping graves
Yawn level with the luminous waves;
But not the riches there that lie
In each idol's diamond eye —
Not the gaily-jewelled dead
Tempt the waters from their bed;
For no ripples curl, alas!
Along that wilderness of glass —
No swellings tell that winds may be
Upon some far-off happier sea —
No heavings hint that winds have been
On seas less hideously serene.
But lo, a stir is in the air!
The wave — there is a movement there!
As if the towers had thrown aside,
In slightly sinking, the dull tide —
As if their tops had feebly given
A void within the filmy Heaven.
The waves have now a redder glow —
The hours are breathing faint and low —
And when, amid no earthly moans,
Down, down that town shall settle hence,
Hell, rising from a thousand thrones,
Shall do it reverence.

88   'The City in the Sea' by Edgar Allan Poe

RIGHT: Portrait of Edgar Allan Poe, unknown maker, late May–early June 1849. © Digital image courtesy of Getty's Open Content Program. The J. Paul Getty Museum, Los Angeles, 84.XT.957

MOBY-DICK
BY HERMAN MELVILLE
(1851)

Today the work of Herman Melville (1819–1891) is regarded as a cornerstone of American literature, but it wasn't until the 20th century, long after his death, that his most famous book, *Moby-Dick; or, The Whale*, was recognised as a Great American Novel. Melville's story was inspired by an account of the albino sperm whale Mocha Dick, published in an 1839 issue of *The Knickerbocker*. In 1840, he joined a whaling voyage aboard the *Acushnet*, which would provide him with the material for a string of maritime novels, starting with *Typee* in 1846.

His sixth novel, *Moby-Dick,* was published in 1851, and is narrated by the sailor Ishmael, whose enigmatic voice leads to long digressions on the history of whaling and cetology, with references to the Bible, Shakespeare, and works of natural history. The monstrous whale lends itself to endless metaphorical interpretations; however, less has been said about the giant squid in Chapter 59. The harpooneer Daggoo first mistakes the squid for the whale, but Captain Ahab, realising what it is, calls off the attack. Ishmael observes the squid's arms, recalls Erik Pontoppidan's description of the kraken in *The Natural History of Norway* (see page 57), and goes on to predict, accurately, that reports of kraken attacks are most likely encounters with giant squid.

LEFT: *Whalers*, Joseph Mallord William Turner, c. 1845.
© Courtesy of The Met Museum, 96.29

Moby-Dick

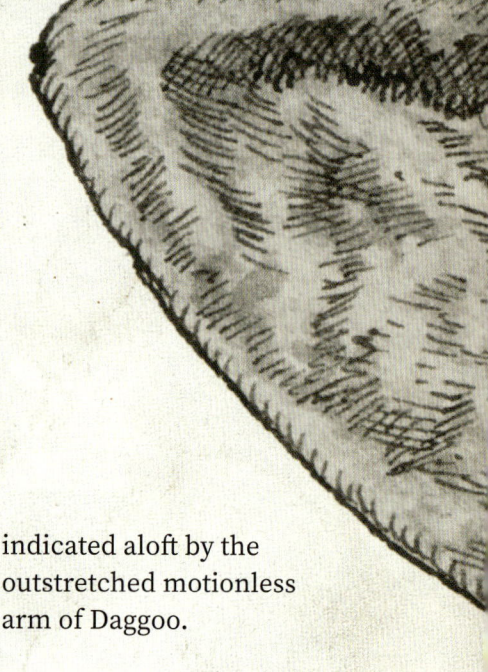

Chapter 59: Squid

Slowly wading through the meadows of brit, the Pequod still held on her way north-eastward towards the island of Java; a gentle air impelling her keel, so that in the surrounding serenity her three tall tapering masts mildly waved to that languid breeze, as three mild palms on a plain. And still, at wide intervals in the silvery night, the lonely, alluring jet would be seen.

But one transparent blue morning, when a stillness almost preternatural spread over the sea, however unattended with any stagnant calm; when the long burnished sun-glade on the waters seemed a golden finger laid across them, enjoining some secrecy; when the slippered waves whispered together as they softly ran on; in this profound hush of the visible sphere a strange spectre was seen by Daggoo from the main-mast-head.

In the distance, a great white mass lazily rose, and rising higher and higher, and disentangling itself from the azure, at last gleamed before our prow like a snow-slide, new slid from the hills. Thus glistening for a moment, as slowly it subsided, and sank. Then once more arose, and silently gleamed. It seemed not a whale; and yet is this Moby-Dick? thought Daggoo. Again the phantom went down, but on re-appearing once more, with a stiletto-like cry that startled every man from his nod, the negro yelled out—"There! there again! there she breaches! right ahead! The White Whale, the White Whale!"

Upon this, the seamen rushed to the yard-arms, as in swarming-time the bees rush to the boughs. Bare-headed in the sultry sun, Ahab stood on the bowsprit, and with one hand pushed far behind in readiness to wave his orders to the helmsman, cast his eager glance in the direction indicated aloft by the outstretched motionless arm of Daggoo.

Whether the flitting attendance of the one still and solitary jet had gradually worked upon Ahab, so that he was now prepared to connect the ideas of mildness and repose with the first sight of the particular whale he pursued; however this was, or whether his eagerness betrayed him; whichever way it might have been, no sooner did he distinctly perceive the white mass, than with a quick intensity he instantly gave orders for lowering.

The four boats were soon on the water; Ahab's in advance, and all swiftly pulling towards their prey. Soon it went down, and while, with oars suspended, we were awaiting its reappearance, lo! in the same spot where it sank, once more it slowly rose.

ABOVE: Drawing of a cuttlefish from *De Animalium Proprietate*, illustrated by Angelo Vergetio, 18C. © Courtesy of the British Library/Bridgeman Images

LEFT: Portrait photograph of Herman Melville, 1860. © Courtesy of Wikimedia Commons

Almost forgetting for the moment all thoughts of Moby-Dick, we now gazed at the most wondrous phenomenon which the secret seas have hitherto revealed to mankind. A vast pulpy mass, furlongs in length and breadth, of a glancing cream-colour, lay floating on the water, innumerable long arms radiating from its centre, and curling and twisting like a nest of anacondas, as if blindly to catch at any hapless object within reach. No perceptible face or front did it have; no conceivable token of either sensation or instinct; but undulated there on the billows, an unearthly, formless, chance-like apparition of life.

As with a low sucking sound it slowly disappeared again, Starbuck still gazing at the agitated waters where it had sunk, with a wild voice exclaimed—"Almost rather had I seen Moby-Dick and fought him, than to have seen thee, thou white ghost!"

"What was it, Sir?" said Flask.

"The great live squid, which, they say, few whale-ships ever beheld, and returned to their ports to tell of it."

But Ahab said nothing; turning his boat, he sailed back to the vessel; the rest as silently following.

Whatever superstitions the sperm whalemen in general have connected with the sight of this object, certain it is, that a glimpse of it being so very unusual, that circumstance has gone far to invest it with portentousness. So rarely is it beheld, that though one and all of them declare it to be the largest animated thing in the ocean, yet very few of them have any but the most vague ideas concerning its true nature and form; notwithstanding, they believe it to furnish to the sperm whale his only food. For though other species of whales find their food above water, and may be seen by man in the act of feeding, the spermaceti whale obtains his whole food in unknown zones below the surface; and only by inference is it that any one can tell of what, precisely, that food consists.

At times, when closely pursued, he will disgorge what are supposed to be the detached arms of the squid; some of them thus exhibited exceeding twenty and thirty feet in length. They fancy that the monster to which these arms belonged ordinarily clings by them to the bed of the ocean; and that the sperm whale, unlike other species, is supplied with teeth in order to attack and tear it.

There seems some ground to imagine that the great Kraken of Bishop Pontoppidan may ultimately resolve itself into Squid. The manner in which the Bishop describes it, as alternately rising and sinking, with some other particulars he narrates, in all this the two correspond. But much abatement is necessary with respect to the incredible bulk he assigns it.

By some naturalists who have vaguely heard rumors of the mysterious creature, here spoken of, it is included among the class of cuttle-fish, to which, indeed, in certain external respects it would seem to belong, but only as the Anak of the tribe.

'LEVIATHAN'
BY CELIA THAXTER
(1861)

The American poet and writer Celia Thaxter (1835–1894) spent her childhood in the Isles of Shoals off the coast of Maine and New Hampshire, where her father was a lighthouse keeper. She married at 16 and moved to Massachusetts. Her first poem, 'Land-Locked', was published in *Atlantic Monthly* in 1861. In 1879 she published the collection *Drift-weed*, which included the enigmatic sea monster poem 'Leviathan'.

In the Hebrew Bible, the Leviathan is a female sea serpent who will bring about the apocalypse with her counterpart, the male desert monster Behemoth. The word is also used to describe the whale that swallows Jonah, and Herman Melville's monstrous *Moby-Dick*. In Thaxter's poem, the Leviathan is a 'placid', 'ethereal, mystic' masculine whale who triumphs against the 'blind rage' of feminine Nature.

Thaxter moved back to the Isles of Shoals in adulthood to run her father's hotel, Appledore House, which became a literary salon, host to writers such as Ralph Waldo Emerson and Nathaniel Hawthorne. The American impressionist painter Childe Hassam was also a frequent visitor, and painted Thaxter on the shore and in the garden on Appledore Island.

LEFT: *This is Leviathan,* illustration from f. 518v of 'The North French Hebrew Miscellany', *c.* 1278–98.
© Courtesy of Wikimedia Commons

'Leviathan'

Betwixt the bleak rock and the barren shore
Rolled miles of hoary waves that hissed with frost,
And from the bitter north with sullen roar
Swept the wild wind, and the wild water tossed.
In the cold sky, hard, pitiless, and drear,
The sun dropped down; but ere the world grew gray,
A sweet, reluctant rose-tint, sad and clear,
Stained icy crags and leagues of leaping spray.
Midway between the lone rock and the shore
A fountain fair sprang skyward suddenly,
And sudden fell, and yet again once more
The column rose, and sank into the sea.
Silent, ethereal, mystic, delicate,
Flushed with delicious glow of fading rose,
It grew and vanished, like some genie great,
Some wild, thin phantom, woven of winter snows.
'T was the foam-fountain of the mighty whale,
Rising each time more far and faint and dim.
All his huge strength against the thundering gale
He set; no hurricane could hinder him!
There came to me a gladness in the sight,
A pleasure in the thought of life so strong,
Daring the elements, and making light
Of winter's wrathful power of wreck and wrong.
I gloried in his triumph o'er the vast
Blind rage of Nature. All her awful force,
The terror of her tempest full she cast
Against him, yet he kept his ponderous course.
For her worst fury he nor stayed nor turned.
'T was joy to think in such tremendous play,
Through the sea's cruelty, all unconcerned.
Leviathan pursued his placid way!

ABOVE: *The South Ledges, Appledore,* Childe Hassam, 1913.
© Courtesy of Smithsonian American Art Museum, Gift of John Gellatly

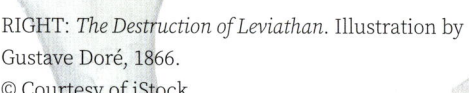

RIGHT: *The Destruction of Leviathan.* Illustration by Gustave Doré, 1866.
© Courtesy of iStock

'Leviathan' by Celia Thaxter

'CALIBAN UPON SETEBOS'
BY ROBERT BROWNING
(1864)

Robert Browning (1812–1889), born in Surrey, England, was a noted poet of the Victorian era and the husband of Elizabeth Barrett Browning (1806–1861), also a poet. Browning is known for his use of the dramatic monologue, a poetic form in which a single character speaks, in a voice distinct from the poet's own voice.

'Caliban upon Setebos' was published in Browning's 1864 collection of dramatic monologues *Dramatis Personae*. It is told from the perspective of Caliban, a bestial creature from William Shakespeare's play *The Tempest* (1611), who is often depicted as a sea monster-human hybrid – referred to as a 'mooncalf' or a 'freckled beast' in the text. He is the son of the now-dead witch Sycorax, who worshipped a god called Setebos. The poem gives voice to the tragic and often abhorrent character, who is enslaved and continually punished by the magician Prospero. It begins with a narrative stanza introducing the scene as Caliban hides in a cave while Prospero and Miranda sleep. Caliban then begins to speak about the natural world Setebos created and the fate of the animals living in it.

'Caliban upon Setebos' was published with the subtitle 'Natural Theology in the Island' – a reference to clergyman William Paley's 1802 Creationist apology *Natural Theology: or, Evidences of the Existence and Attributes of the Deity*, which argued that the intricate beauty of nature must be attributed to an intelligent Creator. Caliban, observing the meaningless cruelty in the world, instead questions Setebos' divine intentions.

LEFT: *Caliban, from 'Twelve Characters from Shakespeare'*, etched by John Hamilton Mortimer, 1775.
© Courtesy of The Met Museum, 62.602.163

'Caliban upon Setebos'

"Thou thoughtest that I was altogether
    such a one as thyself."
(David, Psalms 50.21)

['Will sprawl, now that the heat of day is best,
Flat on his belly in the pit's much mire,
With elbows wide, fists clenched to prop his chin.
And, while he kicks both feet in the cool slush,
And feels about his spine small eft-things course,
Run in and out each arm, and make him laugh:
And while above his head a pompion-plant,
Coating the cave-top as a brow its eye,
Creeps down to touch and tickle hair and beard,
And now a flower drops with a bee inside,
And now a fruit to snap at, catch and crunch,—
He looks out o'er yon sea which sunbeams cross
And recross till they weave a spider-web
(Meshes of fire, some great fish breaks at times)
And talks to his own self, howe'er he please,
Touching that other, whom his dam called God.
Because to talk about Him, vexes—ha,
Could He but know! and time to vex is now,
When talk is safer than in winter-time.
Moreover Prosper and Miranda sleep
In confidence he drudges at their task,
And it is good to cheat the pair, and gibe,
Letting the rank tongue blossom into speech.]
Setebos, Setebos, and Setebos!
'Thinketh, He dwelleth i' the cold o' the moon.
'Thinketh He made it, with the sun to match,
But not the stars; the stars came otherwise;
Only made clouds, winds, meteors, such as that:
Also this isle, what lives and grows thereon,
And snaky sea which rounds and ends the same.
'Thinketh, it came of being ill at ease:
He hated that He cannot change His cold,
Nor cure its ache. 'Hath spied an icy fish
That longed to 'scape the rock-stream where
    she lived,
And thaw herself within the lukewarm brine
O' the lazy sea her stream thrusts far amid,
A crystal spike 'twixt two warm walls of wave;
Only, she ever sickened, found repulse
At the other kind of water, not her life,
(Green-dense and dim-delicious, bred o' the sun)
Flounced back from bliss she was not born
    to breathe,
And in her old bounds buried her despair,
Hating and loving warmth alike: so He.
'Thinketh, He made thereat the sun, this isle,
Trees and the fowls here, beast and creeping thing.
Yon otter, sleek-wet, black, lithe as a leech;
Yon auk, one fire-eye in a ball of foam,
That floats and feeds; a certain badger brown
He hath watched hunt with that slant white-
    wedge eye
By moonlight; and the pie with the long tongue
That pricks deep into oak warts for a worm,
And says a plain word when she finds her prize,
But will not eat the ants; the ants themselves
That build a wall of seeds and settled stalks
About their hole—He made all these and more,
Made all we see, and us, in spite: how else?
He could not, Himself, make a second self

To be His mate; as well have made Himself:
He would not make what He mislikes or slights,
An eyesore to Him, or not worth His pains:
But did, in envy, listlessness or sport,
Make what Himself would fain, in a manner, be—
Weaker in most points, stronger in a few,
Worthy, and yet mere playthings all the while,
Things He admires and mocks too,—that is it.
Because, so brave, so better though they be,
It nothing skills if He begin to plague.
Look, now, I melt a gourd-fruit into mash,
Add honeycomb and pods, I have perceived,
Which bite like finches when they bill and kiss,—
Then, when froth rises bladdery, drink up all,
Quick, quick, till maggots scamper through
    my brain;
Last, throw me on my back i' the seeded thyme,
And wanton, wishing I were born a bird.
Put case, unable to be what I wish,
I yet could make a live bird out of clay:
Would not I take clay, pinch my Caliban
Able to fly?—for, there, see, he hath wings,
And great comb like the hoopoe's to admire,
And there, a sting to do his foes offence,
There, and I will that he begin to live,
Fly to yon rock-top, nip me off the horns
Of grigs high up that make the merry din,
Saucy through their veined wings, and mind me not.
In which feat, if his leg snapped, brittle clay,
And he lay stupid-like,—why, I should laugh;
And if he, spying me, should fall to weep,
Beseech me to be good, repair his wrong,
Bid his poor leg smart less or grow again,—
Well, as the chance were, this might take or else
Not take my fancy: I might hear his cry,
And give the mankin three sound legs for one,
Or pluck the other off, leave him like an egg

ABOVE: Title page of *Dramatis Personæ & Dramatic Romances & Lyrics* by Robert Browning, illustrated by Eleanor Fortescue Brickdale, published by Chatto & Windus, 1909.
© Courtesy of the British Library/Bridgeman Images

And lessoned he was mine and merely clay.
Were this no pleasure, lying in the thyme,
Drinking the mash, with brain become alive,
Making and marring clay at will? So He.
'Thinketh, such shows nor right nor wrong in Him,
Nor kind, nor cruel: He is strong and Lord.
'Am strong myself compared to yonder crabs
That march now from the mountain to the sea;
'Let twenty pass, and stone the twenty-first,
Loving not, hating not, just choosing so.

'Caliban upon Setebos' by Robert Browning

ABOVE: *The Enchanted Island Before the Cell of Prospero. Prospero, Miranda, Caliban and Ariel* (Shakespeare, *The Tempest*, Act I, Scene II). Painted by Henry Fuseli, engraved by Peter Simon, first published 1797; reissued 1852.
© Courtesy of The Met Museum, 42.119.525

'Say, the first straggler that boasts purple spots
Shall join the file, one pincer twisted off;
'Say, this bruised fellow shall receive a worm,
And two worms he whose nippers end in red;
As it likes me each time, I do: so He.
Well then, 'supposeth He is good i' the main,
Placable if His mind and ways were guessed,
But rougher than His handiwork, be sure!
Oh, He hath made things worthier than Himself,
And envieth that, so helped, such things do more
Than He who made them! What consoles but this?
That they, unless through Him, do nought at all,
And must submit: what other use in things?
'Hath cut a pipe of pithless elder-joint
That, blown through, gives exact the scream o' the jay
When from her wing you twitch the feathers blue:
Sound this, and little birds that hate the jay
Flock within stone's throw, glad their foe is hurt:
Put case such pipe could prattle and boast forsooth
"I catch the birds, I am the crafty thing,
I make the cry my maker cannot make

104 'Caliban upon Setebos' by Robert Browning

With his great round mouth; he must blow
    through mine!"
Would not I smash it with my foot? So He.
But wherefore rough, why cold and ill at ease?
Aha, that is a question! Ask, for that,
What knows,—the something over Setebos
That made Him, or He, may be, found and fought,
Worsted, drove off and did to nothing, perchance.
There may be something quiet o'er His head,
Out of His reach, that feels nor joy nor grief,
Since both derive from weakness in some way.
I joy because the quails come; would not joy
Could I bring quails here when I have a mind:
This Quiet, all it hath a mind to, doth.
'Esteemeth stars the outposts of its couch,
But never spends much thought nor care that way.
It may look up, work up,—the worse for those
It works on! 'Careth but for Setebos
The many-handed as a cuttle-fish,
Who, making Himself feared through what
    He does,
Looks up, first, and perceives he cannot soar
To what is quiet and hath happy life;
Next looks down here, and out of very spite
Makes this a bauble-world to ape yon real,
These good things to match those as hips do grapes.
'Tis solace making baubles, ay, and sport.
Himself peeped late, eyed Prosper at his books
Careless and lofty, lord now of the isle:
Vexed, 'stitched a book of broad leaves,
    arrow-shaped,
Wrote thereon, he knows what, prodigious words;
Has peeled a wand and called it by a name;
Weareth at whiles for an enchanter's robe
The eyed skin of a supple oncelot;
And hath an ounce sleeker than youngling mole,
A four-legged serpent he makes cower and couch,
Now snarl, now hold its breath and mind his eye,
And saith she is Miranda and my wife:
'Keeps for his Ariel a tall pouch-bill crane
He bids go wade for fish and straight disgorge;
Also a sea-beast, lumpish, which he snared,
Blinded the eyes of, and brought somewhat tame,
And split its toe-webs, and now pens the drudge
In a hole o' the rock and calls him Caliban;
A bitter heart that bides its time and bites.
'Plays thus at being Prosper in a way,
Taketh his mirth with make-believes: so He.
His dam held that the Quiet made all things
Which Setebos vexed only: 'holds not so.
Who made them weak, meant weakness He
    might vex.
Had He meant other, while His hand was in,
Why not make horny eyes no thorn could prick,
Or plate my scalp with bone against the snow,
Or overscale my flesh 'neath joint and joint
Like an orc's armour? Ay,—so spoil His sport!
He is the One now: only He doth all.
'Saith, He may like, perchance, what profits Him.
Ay, himself loves what does him good; but why?
'Gets good no otherwise. This blinded beast
Loves whoso places flesh-meat on his nose,
But, had he eyes, would want no help, but hate
Or love, just as it liked him: He hath eyes.
Also it pleaseth Setebos to work,
Use all His hands, and exercise much craft,
By no means for the love of what is worked.
'Tasteth, himself, no finer good i' the world
When all goes right, in this safe summer-time,
And he wants little, hungers, aches not much,
Than trying what to do with wit and strength.
'Falls to make something: 'piled yon pile of turfs,
And squared and stuck there squares of soft
    white chalk,

'Caliban upon Setebos' by Robert Browning

ABOVE: 'When Caliban was lazy and neglected his work, Ariel would come slily and pinch him.' Scene from *The Tempest* in *Tales from Shakespeare*, illustrated by Arthur Rackham, 1909.
© Courtesy of Adobe Stock

And, with a fish-tooth, scratched a moon on each,
And set up endwise certain spikes of tree,
And crowned the whole with a sloth's skull a-top,
Found dead i' the woods, too hard for one to kill.
No use at all i' the work, for work's sole sake;
'Shall some day knock it down again: so He.
'Saith He is terrible: watch His feats in proof!
One hurricane will spoil six good months' hope.
He hath a spite against me, that I know,
Just as He favours Prosper, who knows why?
So it is, all the same, as well I find.
'Wove wattles half the winter, fenced them firm
With stone and stake to stop she-tortoises
Crawling to lay their eggs here: well, one wave,
Feeling the foot of Him upon its neck,
Gaped as a snake does, lolled out its large tongue,
And licked the whole labour flat: so much for spite.
'Saw a ball flame down late (yonder it lies)
Where, half an hour before, I slept i' the shade:
Often they scatter sparkles: there is force!
'Dug up a newt He may have envied once
And turned to stone, shut up Inside a stone.
Please Him and hinder this?—What Prosper does?
Aha, if He would tell me how! Not He!
There is the sport: discover how or die!
All need not die, for of the things o' the isle
Some flee afar, some dive, some run up trees;
Those at His mercy,—why, they please Him most
When . . . when . . . well, never try the same
    way twice!
Repeat what act has pleased, He may grow wroth.
You must not know His ways, and play Him off,
Sure of the issue. 'Doth the like himself:
'Spareth a squirrel that it nothing fears
But steals the nut from underneath my thumb,
And when I threat, bites stoutly in defence:
'Spareth an urchin that contrariwise,
Curls up into a ball, pretending death
For fright at my approach: the two ways please.
But what would move my choler more than this,
That either creature counted on its life
To-morrow and next day and all days to come,
Saying, forsooth, in the inmost of its heart,
"Because he did so yesterday with me,
And otherwise with such another brute,
So must he do henceforth and always."—Ay?
'Would teach the reasoning couple what
    "must" means!
'Doth as he likes, or wherefore Lord? So He.

106 'Caliban upon Setebos' by Robert Browning

'Conceiveth all things will continue thus,
And we shall have to live in fear of Him
So long as He lives, keeps His strength: no change,
If He have done His best, make no new world
To please Him more, so leave off watching this,—
If He surprise not even the Quiet's self
Some strange day,—or, suppose, grow into it
As grubs grow butterflies: else, here are we,
And there is He, and nowhere help at all.
'Believeth with the life, the pain shall stop.
His dam held different, that after death
He both plagued enemies and feasted friends:
Idly! He doth His worst in this our life,
Giving just respite lest we die through pain,
Saving last pain for worst,—with which, an end.
Meanwhile, the best way to escape His ire
Is, not to seem too happy. 'Sees, himself,
Yonder two flies, with purple films and pink,
Bask on the pompion-bell above: kills both.
'Sees two black painful beetles roll their ball
On head and tail as if to save their lives:
Moves them the stick away they strive to clear.
Even so, 'would have Him misconceive, suppose
This Caliban strives hard and ails no less,
And always, above all else, envies Him;
Wherefore he mainly dances on dark nights,
Moans in the sun, gets under holes to laugh,
And never speaks his mind save housed as now:
Outside, 'groans, curses. If He caught me here,
O'erheard this speech, and asked "What
    chucklest at?"
'Would, to appease Him, cut a finger off,
Or of my three kid yearlings burn the best,
Or let the toothsome apples rot on tree,
Or push my tame beast for the orc to taste:
While myself lit a fire, and made a song
And sung it, *"What I hate, be consecrate*

ABOVE: Engraving of Robert Browning by George Cook, 1835.
© Courtesy of Wikimedia Commons

*To celebrate Thee and Thy state, no mate
For Thee; what see for envy in poor me?"*
Hoping the while, since evils sometimes mend,
Warts rub away and sores are cured with slime,
That some strange day, will either the Quiet catch
And conquer Setebos, or likelier He
Decrepit may doze, doze, as good as die.
[What, what? A curtain o'er the world at once!
Crickets stop hissing: not a bird—or, yes,
There scuds His raven that has told Him all!
It was fool's play, this prattling! Ha! The wind
Shoulders the pillared dust, death's house o'
    the move,
And fast invading fires begin! White blaze—
A tree's head snaps—and there, there, there,
    there, there,
His thunder follows! Fool to gibe at Him!
Lo! 'Lieth flat and loveth Setebos!
'Maketh his teeth meet through his upper lip,
Will let those quails fly, will not eat this month
One little mess of whelks, so he may 'scape!]

TOILERS OF THE SEA

BY VICTOR HUGO
(1866)

Victor-Marie Hugo (1802–1885) was a French politician and prolific writer, best known for *The Hunchback of Notre-Dame* (1831) and *Les Misérables* (1862). Hugo was born in Besançon in eastern France and lived much of his adult life in Paris. In 1848, he was elected to the National Assembly of the Second French Republic as a conservative but soon broke with his party due to his progressive ideals, notably his opposition to the death penalty.

During the reign of Napoleon III, whom Hugo despised, Hugo lived in Brussels, then on the island of Guernsey in the English Channel. He remained there until Napoleon III was deposed in 1870. His 1866 novel *Les Travailleurs de la Mer* is set on Guernsey. In the story, Gilliatt wrestles a giant octopus, which Hugo refers to using Guernésiais French word for octopus, *pieuvre*, instead of the French *poulpe*, popularising the use of this word on the French mainland. The English translation, *Toilers of the Sea*, uses the term 'devil-fish'. In this excerpt from the novel, the protagonist Gilliatt is attempting to salvage a valuable steam engine from the wreck of the *Durande* when he is attacked by the *pieuvre*.

LEFT: Drawing of an octopus by Victor Hugo, 1866.
© Courtesy of Wikimedia Commons

Toilers of the Sea
Part II, Book IV: Pitfalls in the way

I: HE WHO IS HUNGRY IS NOT ALONE

When he awakened he was hungry.

The sea was growing calmer. But there was still a heavy swell, which made his departure, for the present at least, impossible. The day, too, was far advanced. For the sloop with its burden to get to Guernsey before midnight, it was necessary to start in the morning.

Although pressed by hunger, Gilliatt began by stripping himself, the only means of getting warmth. His clothing was saturated by the storm, but the rain had washed out the sea-water, which rendered it possible to dry them.

He kept nothing on but his trousers, which he turned up nearly to the knees.

His overcoat, jacket, overalls, and sheepskin he spread out and fixed with large round stones here and there.

Then he thought of eating.

He had recourse to his knife, which he was careful to sharpen, and to keep always in good condition; and he detached from the rocks a few limpets, similar in kind to the *clonisses* of the Mediterranean. It is well known that these are eaten raw: but after so many labours, so various and so rude, the pittance was meagre. His biscuit was gone; but of water he had now abundance.

He took advantage of the receding tide to wander among the rocks in search of crayfish. There was extent enough of rock to hope for a successful search.

But he had not reflected that he could do nothing with these without fire to cook them. If he had taken the trouble to go to his store-cavern, he would have found it inundated with the rain. His wood and coal were drowned, and of his store of tow, which served him for tinder, there was not a fibre which was not saturated. No means remained of lighting a fire.

For the rest, his blower was completely disorganised. The screen of the hearth of his forge was broken down; the storm had sacked and devastated his workshop. With what tools and apparatus had escaped the general wreck, he could still have done carpentry work; but he could not have accomplished any of the labours of the smith. Gilliatt, however, never thought of his workshop for a moment.

Drawn in another direction by the pangs of hunger, he had pursued without much reflection his search for food. He wandered, not in the gorge of the rocks, but outside among the smaller breakers. It was there that the *Durande*, ten weeks previously, had first struck upon the sunken reef.

For the search that Gilliatt was prosecuting, this part was more favourable than the interior. At low water the crabs are accustomed to crawl out into the air. They seem to like to warm themselves in the sun, where they swarm sometimes to the disgust of loiterers, who recognise in these creatures, with their awkward sidelong gait, climbing clumsily from crack to crack the lower stages of the rocks like the steps of a

ABOVE: Victor Hugo on a 1959 French banknote.
© Courtesy of Wikimedia Commons

staircase, a sort of sea vermin.

For two months Gilliatt had lived upon these vermin of the sea.

On this day, however, the crayfish and crabs were both wanting. The tempest had driven them into their solitary retreats; and they had not yet mustered courage to venture abroad. Gilliatt held his open knife in his hand, and from time to time scraped a cockle from under the bunches of seaweed, which he ate while still walking.

He could not have been far from the very spot where Sieur Clubin had perished.

As Gilliatt was determining to content himself with the sea-urchins and the *châtaignes de mer*, a little clattering noise at his feet aroused his attention. A large crab, startled by his approach, had just dropped into a pool. The water was shallow, and he did not lose sight of it.

He chased the crab along the base of the rock; the crab moved fast.

Suddenly it was gone.

It had buried itself in some crevice under the rock.

Gilliatt clutched the projections of the rock, and stretched out to observe where it shelved away under the water.

As he suspected, there was an opening there in which the creature had evidently taken refuge. It was more than a crevice; it was a kind of porch.

The sea entered beneath it, but was not deep. The bottom was visible, covered with large pebbles. The pebbles were green and clothed with *confervæ*, indicating that they were never dry. They were like the tops of a number of heads of infants, covered with a kind of green hair.

Holding his knife between his teeth, Gilliatt descended, by the help of feet and hands, from the upper part of the escarpment, and leaped into the water. It reached almost to his shoulders.

He made his way through the porch, and found himself in a blind passage, with a roof in the form of a rude arch over his head. The walls were

Toilers of the Sea by Victor Hugo 111

polished and slippery. The crab was nowhere visible. He gained his feet and advanced in daylight growing fainter, so that he began to lose the power to distinguish objects.

At about fifteen paces the vaulted roof ended overhead. He had penetrated beyond the blind passage. There was here more space, and consequently more daylight. The pupils of his eyes, moreover, had dilated; he could see pretty clearly. He was taken by surprise.

He had made his way again into the singular cavern which he had visited in the previous month. The only difference was that he had entered by the way of the sea.

It was through the submarine arch, that he had remarked before, that he had just entered. At certain low tides it was accessible.

His eyes became more accustomed to the place. His vision became clearer and clearer. He was astonished. He found himself again in that extraordinary palace of shadows; saw again before his eyes that vaulted roof, those columns, those purple and blood-like stains, that vegetation rich with gems, and at the farther end, that crypt or sanctuary, and that altar-like stone. He took little notice of these details, but their impression was in his mind, and he saw that the place was unchanged.

He observed before him, at a certain height in the wall, the crevice through which he had penetrated the first time, and which, from the point where he now stood, appeared inaccessible.

Near the moulded arch, he remarked those low dark grottoes, a sort of caves within a cavern, which he had already observed from a distance. He now stood nearer to them. The entrance to the nearest to him was out of the water, and easily approachable. Nearer still than this recess he noticed, above the level of the water, and within reach of his hand, a horizontal fissure. It seemed to him probable that the crab had taken refuge there, and he plunged his hand in as far as he was able, and groped about in that dusky aperture.

Suddenly he felt himself seized by the arm. A strange indescribable horror thrilled through him.

Some living thing, thin, rough, flat, cold, slimy, had twisted itself round his naked arm, in the dark depth below. It crept upward towards his chest. Its pressure was like a tightening cord, its steady persistence like that of a screw. In less than a moment some mysterious spiral form had passed round his wrist and elbow, and had reached his shoulder. A sharp point penetrated beneath the armpit.

Gilliatt recoiled; but he had scarcely power to move! He was, as it were, nailed to the place. With his left hand, which was disengaged, he seized his knife, which he still held between his teeth, and with that hand, holding the knife, he supported himself against the rocks, while he made a desperate effort to withdraw his arm. He succeeded only in disturbing his persecutor, which wound itself still tighter. It was supple as leather, strong as steel, cold as night.

A second form, sharp, elongated, and narrow, issued out of the crevice, like a tongue out of monstrous jaws. It seemed to lick his naked body. Then suddenly stretching out, it became longer and thinner, as it crept over his skin, and wound itself round him. At the same time a terrible sense of pain, comparable to nothing he had ever known, compelled all his muscles to contract. He felt upon his skin a number of flat rounded points. It seemed as if innumerable suckers had fastened to his flesh and were about to drink his blood.

A third long undulating shape issued from the hole in the rock; seemed to feel its way about his body; lashed round his ribs like a cord, and fixed itself there.

Agony when at its height is mute. Gilliatt uttered no cry. There was sufficient light for him to see the repulsive forms which had entangled themselves about him. A fourth ligature, but this one swift as an arrow, darted towards his stomach, and wound around him there.

It was impossible to sever or tear away the slimy bands which were twisted tightly round his body, and were adhering by a number of points. Each of the points was the focus of frightful and singular pangs. It was as if numberless small mouths were devouring him at the same time.

A fifth long, slimy, riband-shaped strip issued from the hole. It passed over the others, and wound itself tightly around his chest. The compression increased his sufferings. He could scarcely breathe.

These living thongs were pointed at their extremities, but broadened like a blade of a sword towards its hilt. All belonged evidently to the same centre. They crept and glided about him; he felt the strange points of pressure, which seemed to him like mouths, change their places from time to time.

Suddenly a large, round, flattened, glutinous mass issued from beneath the crevice. It was the centre; the five thongs were attached to it like spokes to the nave of a wheel. On the opposite side of this disgusting monster appeared the commencement of three other tentacles, the ends of which remained under the rock. In the middle of this slimy mass appeared two eyes.

The eyes were fixed on Gilliatt.

He recognised the Devil-Fish.

II: THE MONSTER

It is difficult for those who have not seen it to believe in the existence of the devil-fish.

Compared to this creature, the ancient hydras are insignificant.

At times we are tempted to imagine that the vague forms which float in our dreams may encounter in the realm of the Possible attractive forces, having power to fix their lineaments, and shape living beings, out of these creatures of our slumbers. The Unknown has power over these strange visions, and out of them composes monsters. Orpheus, Homer, and Hesiod imagined only the Chimera: Providence has created this terrible creature of the sea.

RIGHT: Illustration of an octopus from *Aquatilium Animalium Historiae* (The History of Aquatic Animals) by Ippolito Salviani, 1554.
© Courtesy of Wikimedia Commons

Creation abounds in monstrous forms of life. The wherefore of this perplexes and affrights the religious thinker.

If terror were the object of its creation, nothing could be imagined more perfect than the devil-fish.

The whale has enormous bulk, the devil-fish is comparatively small; the jararaca makes a hissing noise, the devil-fish is mute; the rhinoceros has a horn, the devil-fish has none; the scorpion has a dart, the devil-fish has no dart; the shark has sharp fins, the devil-fish has no fins; the vespertilio-bat has wings with claws, the devil-fish has no wings; the porcupine has his spines, the devil-fish has no spines; the sword-fish has his sword, the devil-fish has none; the torpedo has its electric spark, the devil-fish has none; the toad has its poison, the devil-fish has none; the viper has its venom, the devil-fish has no venom; the lion has its talons, the devil-fish has no talons; the griffon has its beak, the devil-fish has no beak; the crocodile has its jaws, the devil-fish has no teeth.

The devil-fish has no muscular organisation, no menacing cry, no breastplate, no horn, no dart, no claw, no tail with which to hold or bruise; no cutting fins, or wings with nails, no prickles, no sword, no electric discharge, no poison, no talons, no beak, no teeth. Yet he is of all creatures the most formidably armed.

What, then, is the devil-fish? It is the sea vampire.

The swimmer who, attracted by the beauty of the spot, ventures among breakers in the open sea, where the still waters hide the splendours of the deep, or in the hollows of unfrequented rocks, in unknown caverns abounding in sea plants, testacea, and crustacea, under the deep portals of the ocean, runs the risk of meeting it. If that fate should be yours, be not curious, but fly. The intruder enters there dazzled; but quits the spot in terror.

This frightful apparition, which is always possible among the rocks in the open sea, is a greyish form which undulates in the water. It is of the thickness of a man's arm, and in length nearly five feet. Its outline is ragged. Its form resembles an umbrella closed, and without handle. This irregular mass advances slowly towards you. Suddenly it opens, and eight radii issue abruptly from around a face with two eyes. These radii are alive: their undulation is like lambent flames; they resemble, when opened, the spokes of a wheel, of four or five feet in diameter. A terrible expansion! It springs upon its prey.

The devil-fish harpoons its victim.

It winds around the sufferer, covering and entangling him in its long folds. Underneath it is yellow; above, a dull, earthy hue: nothing could render that inexplicable shade dust coloured. Its form is spider-like, but its tints are like those of the chamelion. When irritated it becomes violet. Its most horrible characteristic is its softness.

Its folds strangle, its contact paralyses.

It has an aspect like gangrened or scabrous flesh. It is a monstrous embodiment of disease.

It adheres closely to its prey, and cannot be torn away; a fact which is due to its power of exhausting air. The eight antennæ, large at their roots, diminish gradually, and end in needle-like points. Underneath each of these feelers range two rows of pustules, decreasing in size, the largest ones near the head, the smaller at the extremities. Each row contains twenty-five of these. There are, therefore, fifty pustules to each feeler, and the creature possesses in the whole four hundred. These pustules are capable of acting like cupping-glasses. They are cartilaginous substances, cylindrical, horny, and livid. Upon the large species they diminish gradually from the diameter of a five-franc piece to the size of a split pea. These small tubes can be thrust out and withdrawn by the animal at will. They are capable of piercing to a depth of more than an inch.

This sucking apparatus has all the regularity and delicacy of a key-board. It stands forth at one moment and disappears the next. The most perfect sensitiveness cannot equal the contractibility of these suckers; always proportioned to the internal movement of the animal, and its exterior circumstances. The monster is endowed with the qualities of the sensitive plant.

This animal is the same as those which mariners call Poulps; which science designates Cephalopteræ, and which ancient legends call Krakens. It is the English sailors who call them "Devil-fish," and sometimes Bloodsuckers. In the Channel Islands they are called *pieuvres*.

They are rare at Guernsey, very small at Jersey; but near the island of Sark are numerous as well as very large.

An engraving in Sonnini's edition of Buffon represents a Cephaloptera crushing a frigate. Denis Montfort, in fact, considers the Poulp, or Octopod, of high latitudes, strong enough to destroy a ship. Bory Saint Vincent doubts this; but he shows that in our regions they will attack men. Near Brecq-Hou, in Sark, they show a cave where a devil-fish a few years since seized and drowned a lobster-fisher. Peron and Lamarck are in error in their belief that the "poulp" having no fins cannot swim. He who writes these lines has seen with his own eyes, at Sark, in the cavern called the Boutiques, a *pieuvre* swimming and pursuing a bather. When captured and killed, this specimen was found to be four English feet broad, and it was possible to count its four hundred suckers. The monster thrust them out convulsively in the agony of death.

According to Denis Montfort, one of those observers whose marvellous intuition sinks or raises them to the level of magicians, the poulp is almost endowed with the passions of man: it has its hatreds. In fact, in the Absolute to be hideous is to hate.

Hideousness struggles under the natural law of elimination, which necessarily renders it hostile.

When swimming, the devil-fish rests, so to speak, in its sheath. It swims with all its parts drawn close. It may be likened to a sleeve sewn up with a closed fist within. The protuberance, which is the head, pushes the water aside and advances with a vague undulatory movement. Its two eyes, though large, are indistinct, being of the colour of the water.

When in ambush, or seeking its prey, it retires into itself, grows smaller and condenses itself. It is then scarcely distinguishable in the submarine twilight.

At such times, it looks like a mere ripple in the water. It resembles anything except a living creature.

The devil-fish is crafty. When its victim is unsuspicious, it opens suddenly.

A glutinous mass, endowed with a malignant will, what can be more horrible?

116   Toilers of the Sea by Victor Hugo

It is in the most beautiful azure depths of the limpid water that this hideous, voracious polyp delights. It always conceals itself, a fact which increases its terrible associations. When they are seen, it is almost invariably after they have been captured.

At night, however, and particularly in the hot season, it becomes phosphorescent. These horrible creatures have their passions; their submarine nuptials. Then it adorns itself, burns and illumines; and from the height of some rock, it may be seen in the deep obscurity of the waves below, expanding with a pale irradiation—a spectral sun.

The devil-fish not only swims, it walks. It is partly fish, partly reptile. It crawls upon the bed of the sea. At these times, it makes use of its eight feelers, and creeps along in the fashion of a species of swift-moving caterpillar.

It has no blood, no bones, no flesh. It is soft and flabby; a skin with nothing inside. Its eight tentacles may be turned inside out like the fingers of a glove.

It has a single orifice in the centre of its radii, which appears at first to be neither the vent nor the mouth. It is, in fact, both one and the other. The orifice performs a double function. The entire creature is cold.

The jelly-fish of the Mediterranean is repulsive. Contact with that animated gelatinous substance which envelopes the bather, in which the hands sink, and the nails scratch ineffectively; which can be torn without killing it, and which can be plucked off without entirely removing it—that fluid and yet tenacious creature which slips through the fingers, is disgusting; but no horror can equal the sudden apparition of the devil-fish, that Medusa with its eight serpents.

No grasp is like the sudden strain of the cephaloptera.

It is with the sucking apparatus that it attacks. The victim is oppressed by a vacuum drawing at numberless points: it is not a clawing or a biting, but an indescribable scarification. A tearing of the flesh is terrible, but less terrible than a sucking of the blood. Claws are harmless compared with the horrible action of these natural air-cups. The talons of the wild beast enter into your flesh; but with the cephaloptera it is you who enter into the creature. The muscles swell, the fibres of the body are contorted, the skin cracks under the loathsome oppression, the blood spurts out and mingles horribly with the lymph of the monster, which clings to its victim by innumerable hideous mouths. The hydra incorporates itself with the man; the man becomes one with the hydra. The spectre lies upon you: the tiger can only devour you; the devil-fish, horrible, sucks your life-blood away. He draws you to him, and into himself; while bound down, glued to the ground, powerless, you feel yourself gradually emptied into this horrible pouch, which is the monster itself.

These strange animals, Science, in accordance with its habit of excessive caution even in the face of facts, at first rejects as fabulous; then she decides to observe them; then she dissects, classifies, catalogues, and labels; then procures specimens, and exhibits them in glass cases in museums. They enter then into her nomenclature; are designated mollusks, invertebrata, radiata: she determines their position in the animal world a little above the calamaries, a little below the cuttle-fish; she finds for these hydras of the sea an analogous

Toilers of the Sea by Victor Hugo

ABOVE: 'Gamochonia–Trichterkraken' from *Kunstformen der Natur*, by Ernst Haeckel, 1904. © Courtesy of Wikimedia Commons

creature in fresh water called the argyronecte: she divides them into great, medium, and small kinds; she admits more readily the existence of the small than of the large species, which is, however, the tendency of science in all countries, for she is by nature more microscopic than telescopic. She regards them from the point of view of their construction, and calls them Cephaloptera; counts their antennæ, and calls them Octopedes. This done, she leaves them. Where science drops them, philosophy takes them up.

Philosophy in her turn studies these creatures. She goes both less far and further. She does not dissect, but meditate. Where the scalpel has laboured, she plunges the hypothesis. She seeks the final cause. Eternal perplexity of the thinker. These creatures disturb his ideas of the Creator. They are hideous surprises. They are the death's-head at the feast of contemplation. The philosopher determines their characteristics in dread. They are the concrete forms of evil. What attitude can he take towards this treason of creation against herself? To whom can he look for the solution of these riddles? The Possible is a terrible matrix. Monsters are mysteries in their concrete form. Portions of shade issue from the mass, and something within detaches itself, rolls, floats,

118　Toilers of the Sea by Victor Hugo

condenses, borrows elements from the ambient darkness, becomes subject to unknown polarisations, assumes a kind of life, furnishes itself with some unimagined form from the obscurity, and with some terrible spirit from the miasma, and wanders ghostlike among living things. It is as if night itself assumed the forms of animals. But for what good? with what object? Thus we come again to the eternal questioning.

These animals are indeed phantoms as much as monsters. They are proved and yet improbable. Their fate is to exist in spite of *à priori* reasonings. They are the amphibia of the shore which separates life from death. Their unreality makes their existence puzzling. They touch the frontier of man's domain and people the region of chimeras. We deny the possibility of the vampire, and the cephaloptera appears. Their swarming is a certainty which disconcerts our confidence. Optimism, which is nevertheless in the right, becomes silenced in their presence. They form the visible extremity of the dark circles. They mark the transition of our reality into another. They seem to belong to that commencement of terrible life which the dreamer sees confusedly through the loophole of the night.

That multiplication of monsters, first in the Invisible, then in the Possible, has been suspected, perhaps perceived by magi and philosophers in their austere ecstasies and profound contemplations. Hence the conjecture of a material hell. The demon is simply the invisible tiger. The wild beast which devours souls has been presented to the eyes of human beings by St. John, and by Dante in his vision of Hell.

If, in truth, the invisible circles of creation continue indefinitely, if after one there is yet another, and so forth in illimitable progression; if that chain, which for our part we are resolved to doubt, really exist, the cephaloptera at one extremity proves Satan at the other. It is certain that the wrongdoer at one end proves the existence of wrong at the other.

Every malignant creature, like every perverted intelligence, is a sphinx. A terrible sphinx propounding a terrible riddle; the riddle of the existence of Evil.

It is this perfection of evil which has sometimes sufficed to incline powerful intellects to a faith in the duality of the Deity, towards that terrible bifrons of the Manichæans.

A piece of silk stolen during the last war from the palace of the Emperor of China represents a shark eating a crocodile, who is eating a serpent, who is devouring an eagle, who is preying on a swallow, who in his turn is eating a caterpillar.

All nature which is under our observation is thus alternately devouring and devoured. The prey prey on each other.

Learned men, however, who are also philosophers, and therefore optimists in their view of creation, find, or believe they find, an explanation. Among others, Bonnet of Geneva, that mysterious exact thinker, who was opposed to Buffon, as in later times Geoffrey St. Hilaire has been to Cuvier, was struck with the idea of the final object. His notions may be summed up thus: universal death necessitates universal sepulture; the devourers are the sextons of the system of nature. All created things enter into and form the elements of other. To decay is to nourish. Such is the terrible law from which not even man himself escapes.

In our world of twilight this fatal order of things produces monsters. You ask for what purpose. We find the solution here.

But *is* this the solution? Is this the answer to our questionings? And if so, why not some different order of things? Thus the question returns.

Let us live: be it so.

But let us endeavour that death shall be progress. Let us aspire to an existence in which these mysteries shall be made clear. Let us follow that conscience which leads us thither.

For let us never forget that the highest is only attained through the high.

ANOTHER KIND OF SEA-COMBAT

Such was the creature in whose power Gilliatt had fallen for some minutes.

The monster was the inhabitant of the grotto; the terrible genii of the place. A kind of sombre demon of the water.

All the splendours of the cavern existed for it alone.

On the day of the previous month when Gilliatt had first penetrated into the grotto, the dark outline, vaguely perceived by him in the ripples of the secret waters, was this monster. It was here in its home.

When entering for the second time into the cavern in pursuit of the crab, he had observed the crevice in which he supposed that the crab had taken refuge, the *pieuvre* was there lying in wait for prey.

Is it possible to imagine that secret ambush?

No bird would brood, no egg would burst to life, no flower would dare to open, no breast to give milk, no heart to love, no spirit to soar, under the influence of that apparition of evil watching with sinister patience in the dusk.

Gilliatt had thrust his arm deep into the opening; the monster had snapped at it. It held him fast, as the spider holds the fly.

He was in the water up to his belt; his naked feet clutching the slippery roundness of the huge stones at the bottom; his right arm bound and rendered powerless by the flat coils of the long tentacles of the creature, and his body almost hidden under the folds and cross folds of this horrible bandage.

Of the eight arms of the devil-fish three adhered to the rock, while five encircled Gilliatt. In this way, clinging to the granite on the one hand, and with the other to its human prey, it enchained him to the rock. Two hundred and fifty suckers were upon him, tormenting him with agony and loathing. He was grasped by gigantic hands, the fingers of which were each nearly a yard long, and furnished inside with living blisters eating into the flesh.

As we have said, it is impossible to tear oneself from the folds of the devil-fish. The attempt ends only in a firmer grasp. The monster clings with more determined force. Its effort increases with that of its victim; every struggle produces a tightening of its ligatures.

Gilliatt had but one resource, his knife.

His left hand only was free; but the reader knows with what power he could use it. It might have been said that he had two right hands.

His open knife was in his hand.

The antenna of the devil-fish cannot be cut; it is a leathery substance impossible to divide with the knife, it slips under the edge; its position in attack also is such that to cut it would be to wound the victim's own flesh.

The creature is formidable, but there is a way of resisting it. The fishermen of Sark know this, as does any one who has seen them execute certain movements in the sea. The porpoises know it also; they have a way of biting the cuttle-fish which decapitates it. Hence the frequent sight on the sea of pen-fish, poulps, and cuttle-fish without heads.

The cephaloptera, in fact, is only vulnerable through the head.

Gilliatt was not ignorant of this fact.

He had never seen a devil-fish of this size. His first encounter was with one of the larger species. Another would have been powerless with terror.

With the devil-fish, as with a furious bull, there is a certain moment in the conflict which must be seized. It is the instant when the bull lowers the neck; it is the instant when the devil-fish advances its head. The movement is rapid. He who loses that moment is destroyed.

The things we have described occupied only a few moments. Gilliatt, however, felt the increasing power of its innumerable suckers.

The monster is cunning; it tries first to stupefy its prey. It seizes and then pauses awhile.

Gilliatt grasped his knife; the sucking increased.

He looked at the monster, which seemed to look at him.

Suddenly it loosened from the rock its sixth antenna, and darting it at him, seized him by the left arm.

At the same moment it advanced its head with a violent movement. In one second more its mouth would have fastened on his breast. Bleeding in the sides, and with his two arms entangled, he would have been a dead man.

But Gillian was watchful. He avoided the antenna, and at the moment when the monster darted forward to fasten on his breast, he struck it with the knife clenched in his left hand. There were two convulsions in opposite directions; that of the devil-fish and that of its prey. The movement was rapid as a double flash of lightnings.

He had plunged the blade of his knife into the flat slimy substance, and by a rapid movement, like the flourish of a whip in the air, describing a circle round the two eyes, he wrenched the head off as a man would draw a tooth.

The struggle was ended. The folds relaxed. The monster dropped away, like the slow detaching of bands. The four hundred suckers, deprived of their sustaining power, dropped at once from the man and the rock. The mass sank to the bottom of the water.

Breathless with the struggle, Gilliatt could perceive upon the stones at his feet two shapeless, slimy heaps, the head on one side, the remainder of the monster on the other.

Fearing, nevertheless, some convulsive return of his agony, he recoiled to avoid the reach of the dreaded tentacles.

But the monster was quite dead.

Gilliatt closed his knife.

Twenty Thousand Leagues Under the Sea

by Jules Verne
(1871)

The French novelist Jules Verne (1828–1905) is the author of the *Voyages extraordinaires*, a series of novels and adventure stories within the *Merveilleux scientifique* (literally: scientific marvellous) genre of popular fiction. *Vingt Mille Lieues Sous Les Mers* (Twenty Thousand Leagues Under the Sea) was serialised in *Magasin d'éducation et de recreation* from 1869 to 1870 and published as a novel in 1871. The story follows the journey of Captain Nemo and his submarine the *Nautilus*, which is capable of sustaining life almost indefinitely. It is narrated by Professor Pierre Aronnax, a French scientist, who is a captive of Captain Nemo, along with his Flemish servant Conseil and the Canadian harpooner Ned Land.

Verne took the name for his submarine from the early French naval vessel. He was further inspired by the submarine *Plongeur*, which he saw at the 1867 *Exhibition Universelle* world's fair in Paris. This exhibition also led Verne to integrate the relatively new concept of electricity into his designs for the *Nautilus*, which draws power from electrolytes extracted from seawater. Verne's main oceanographic resource was MF Maury's 1855 *The Physical Geography of the Sea*. Verne's blend of speculative technology and real natural philosophy has led to the classification of much of his work as science fiction, although the term was not widely used until later.

LEFT: 'A walk under the waters', from *Twenty Thousand Leagues Under the Sea*, illustrated by Alphonse de Neuville and Édouard Riou, 1871.
© Courtesy of Wikimedia Commons

Twenty Thousand Leagues Under the Sea
Part II, Chapter XVIII, The Poulps

For several days the *Nautilus* kept off from the American coast. Evidently it did not wish to risk the tides of the Gulf of Mexico or of the sea of the Antilles. April 16th, we sighted Martinique and Guadaloupe from a distance of about thirty miles. I saw their tall peaks for an instant. The Canadian, who counted on carrying out his projects in the Gulf, by either landing or hailing one of the numerous boats that coast from one island to another, was quite disheartened. Flight would have been quite practicable, if Ned Land had been able to take possession of the boat without the Captain's knowledge. But in the open sea it could not be thought of. The Canadian, Conseil, and I had a long conversation on this subject. For six months we had been prisoners on board the *Nautilus*. We had travelled 17,000 leagues; and, as Ned Land said, there was no reason why it should come to an end. We could hope nothing from the Captain of the *Nautilus*, but only from ourselves. Besides, for some time past he had become graver, more retired, less sociable. He seemed to shun me. I met him rarely. Formerly he was pleased to explain the submarine marvels to me; now he left me to my studies, and came no more to the saloon. What change had come over him? For what cause? For my part, I did not wish to bury with me my curious and novel studies. I had now the power to write the true book of the sea; and this book, sooner or later, I wished to see daylight. The land nearest us was the archipelago of the Bahamas. There rose high submarine cliffs covered with large weeds. It was about eleven o'clock when Ned Land drew my attention to a formidable pricking, like the sting of an ant, which was produced by means of large seaweeds.

"Well," I said, "these are proper caverns for poulps, and I should not be astonished to see some of these monsters."

"What!" said Conseil; "cuttlefish, real cuttlefish of the cephalopod class?"

"No," I said, "poulps of huge dimensions."

"I will never believe that such animals exist," said Ned.

"Well," said Conseil, with the most serious air in the world, "I remember perfectly to have seen a large vessel drawn under the waves by an octopus's arm."

"You saw that?" said the Canadian.

"Yes, Ned."

"With your own eyes?"

"With my own eyes."

"Where, pray, might that be?"

"At St. Malo," answered Conseil.

"In the port?" said Ned, ironically.

"No; in a church," replied Conseil.

"In a church!" cried the Canadian.

"Yes; friend Ned. In a picture representing the poulp in question."

"Good!" said Ned Land, bursting out laughing.

"He is quite right," I said. "I have heard of this picture; but the subject represented is taken from a legend, and you know what to think of legends in the matter of natural history. Besides, when it is a question of monsters, the imagination is apt to run wild. Not only is it supposed that these poulps can draw down vessels, but a certain Olaus Magnus speaks of an octopus a mile long that is more like an island than an animal. It is also said that the Bishop of Nidros was building an altar on an immense rock. Mass finished, the rock began to walk, and returned to the sea. The rock was a poulp. Another Bishop, Pontoppidan, speaks also of a poulp on which a regiment of cavalry could manœuvre. Lastly, the ancient naturalists speak of monsters whose mouths were like gulfs, and which were too large to pass through the Straits of Gibraltar."

"But how much is true of these stories?" asked Conseil.

"Nothing, my friends; at least of that which passes the limit of truth to get to fable or legend. Nevertheless, there must be some ground for the imagination of the story-tellers. One cannot deny that poulps and cuttlefish exist of a large species, inferior, however, to the cetaceans. Aristotle has

ABOVE: Portrait of Jules Verne photographed by Whitfield Lock.
© Courtesy of Adobe Stock

LEFT: Frontispiece for *Vingt Mille Lieues Sous Les Mers* in *Voyages extraordinaires*, illustrated by Alphonse de Neuville and Édouard Riou, 1871. © Courtesy of Wikimedia Commons

stated the dimensions of a cuttlefish as five cubits, or nine feet two inches. Our fishermen frequently see some that are more than four feet long. Some skeletons of poulps are preserved in the museums of Trieste and Montpelier, that measure two yards in length. Besides, according to the calculations of some naturalists, one of these animals only six feet long would have tentacles twenty-seven feet long. That would suffice to make a formidable monster."

"Do they fish for them in these days?" asked Ned.

"If they do not fish for them, sailors see them at least. One of my friends, Captain Paul Bos of Havre, has often affirmed that he met one of these monsters of colossal dimensions in the Indian seas. But the most astonishing fact, and which does not permit of the denial of the existence of these gigantic animals, happened some years ago, in 1861."

"What is the fact?" asked Ned Land.

"This is it. In 1861, to the north-east of Teneriffe, very nearly in the same latitude we are in now, the crew of the despatch-boat Alector perceived a monstrous cuttlefish swimming in the waters. Captain Bouguer went near to the animal, and attacked it with harpoon and guns, without much success, for balls and harpoons glided over the soft flesh. After several fruitless attempts the crew tried to pass a slip-knot round the body of the mollusc. The noose slipped as far as the tail fins and there stopped. They tried then to haul it on board, but its weight was so considerable that the tightness of the cord separated the tail from the body, and, deprived of this ornament, he disappeared under the water."

"Indeed! is that a fact?"

"An indisputable fact, my good Ned. They proposed to name this poulp 'Bouguer's cuttlefish.'"

"What length was it?" asked the Canadian.

"Did it not measure about six yards?" said Conseil, who, posted at the window, was examining again the irregular windings of the cliff.

"Precisely," I replied.

"Its head," rejoined Conseil, "was it not crowned with eight tentacles, that beat the water like a nest of serpents?"

"Precisely."

"Had not its eyes, placed at the back of its head, considerable development?"

"Yes, Conseil."

"And was not its mouth like a parrot's beak?"

"Exactly, Conseil."

"Very well! no offence to master," he replied, quietly; "if this is not Bouguer's cuttlefish, it is, at least, one of its brothers."

I looked at Conseil. Ned Land hurried to the window.

"What a horrible beast!" he cried.

I looked in my turn, and could not repress a gesture of disgust. Before my eyes was a horrible monster worthy to figure in the legends of the marvellous. It was an immense cuttlefish, being eight yards long. It swam crossways in the direction of the *Nautilus* with great speed, watching us with its enormous staring green eyes. Its eight arms, or rather feet, fixed to its head, that have given the name of cephalopod to these animals, were twice as long as its body, and were twisted like the furies' hair. One could see the 250 air holes on the inner side of the tentacles. The monster's mouth, a horned beak like a parrot's, opened and shut vertically. Its tongue, a horned substance, furnished with several rows of pointed teeth, came out quivering from this veritable pair of shears. What a freak of nature, a bird's beak on a mollusc! Its spindle-like body formed a fleshy mass that might weigh 4,000 to 5,000 lbs.; the, varying colour changing with great rapidity, according to the irritation of the animal, passed successively from livid grey to reddish brown. What irritated this mollusc? No doubt the presence of the *Nautilus*, more formidable than itself, and on which its suckers or its jaws had no hold. Yet, what monsters these poulps are! what vitality the Creator has given them! what vigour in their movements! and they possess three hearts! Chance had brought us in presence of this cuttlefish, and I did not wish to lose the opportunity of carefully studying this specimen of cephalopods. I overcame the horror that inspired me, and, taking a pencil, began to draw it.

"Perhaps this is the same which the Alector saw," said Conseil.

"No," replied the Canadian; "for this is whole, and the other had lost its tail."

"That is no reason," I replied. "The arms and tails of these animals are re-formed by renewal; and in seven years the tail of Bouguer's cuttlefish has no doubt had time to grow."

By this time other poulps appeared at the port light. I counted seven. They formed a procession after the *Nautilus*, and I heard their beaks gnashing against the iron hull. I continued my work. These monsters kept in the water with such precision that they seemed immovable. Suddenly the *Nautilus* stopped. A shock made it tremble in every plate.

"Have we struck anything?" I asked.

"In any case," replied the Canadian, "we shall be free, for we are floating."

The *Nautilus* was floating, no doubt, but it did not move. A minute passed. Captain Nemo, followed by his lieutenant, entered the drawing-room. I had not seen him for some time. He seemed dull. Without noticing or speaking to us, he went to the panel, looked at the poulps, and said something to his lieutenant. The latter went out. Soon the panels were shut. The ceiling was lighted. I went towards the Captain.

"A curious collection of poulps?" I said.

"Yes, indeed, Mr. Naturalist," he replied; "and we are going to fight them, man to beast."

I looked at him. I thought I had not heard aright.

"Man to beast?" I repeated.

"Yes, sir. The screw is stopped. I think that the horny jaws of one of the cuttlefish is entangled in the blades. That is what prevents our moving."

"What are you going to do?"

"Rise to the surface, and slaughter this vermin."

"A difficult enterprise."

"Yes, indeed. The electric bullets are powerless against the soft flesh, where they do not find resistance enough to go off. But we shall attack them with the hatchet."

"And the harpoon, sir," said the Canadian, "if you do not refuse my help."

"I will accept it, Master Land."

"We will follow you," I said, and, following Captain Nemo, we went towards the central staircase.

There, about ten men with boarding-hatchets were ready for the attack. Conseil and I took two hatchets; Ned Land seized a harpoon. The *Nautilus* had then risen to the surface. One of the sailors, posted on the top ladderstep, unscrewed the bolts of the panels. But hardly were the screws loosed, when the panel rose with great violence, evidently drawn by the suckers of a poulp's arm. Immediately one of these arms slid like a serpent down the opening and twenty others were above. With one blow of the axe, Captain Nemo cut this formidable tentacle, that slid wriggling down the ladder. Just as we were pressing one on the other to reach the platform, two other arms, lashing the air, came down on the seaman placed before Captain Nemo, and lifted him up with irresistible power. Captain Nemo uttered a cry, and rushed out. We hurried after him.

What a scene! The unhappy man, seized by the tentacle and fixed to the suckers, was balanced in the air at the caprice of this enormous trunk. He rattled in his throat, he was stifled, he cried, "Help! help!" These words, spoken in French, startled me! I had a fellow-countryman on board, perhaps several! That heart-rending cry! I shall hear it all my life. The unfortunate

man was lost. Who could rescue him from that powerful pressure? However, Captain Nemo had rushed to the poulp, and with one blow of the axe had cut through one arm. His lieutenant struggled furiously against other monsters that crept on the flanks of the *Nautilus*. The crew fought with their axes. The Canadian, Conseil, and I buried our weapons in the fleshy masses; a strong smell of musk penetrated the atmosphere. It was horrible!

For one instant, I thought the unhappy man, entangled with the poulp, would be torn from its powerful suction. Seven of the eight arms had been cut off. One only wriggled in the air, brandishing the victim like a feather. But just as Captain Nemo and his lieutenant threw themselves on it, the animal ejected a stream of black liquid. We were blinded with it. When the cloud dispersed, the cuttlefish had disappeared, and my unfortunate countryman with it. Ten or twelve poulps now invaded the platform and sides of the *Nautilus*. We rolled pell-mell into the midst of this nest of serpents, that wriggled on the platform in the waves of blood and ink. It seemed as though these slimy tentacles sprang up like the hydra's heads. Ned Land's harpoon, at each stroke, was plunged into the staring eyes of the cuttle fish. But my bold companion was suddenly overturned by the tentacles of a monster he had not been able to avoid.

Ah! how my heart beat with emotion and horror! The formidable beak of a cuttlefish was open over Ned Land. The unhappy man would be cut in two. I rushed to his succour. But Captain Nemo was before me; his axe disappeared between the two enormous jaws, and, miraculously saved, the Canadian, rising, plunged his harpoon deep into the triple heart of the poulp.

"I owed myself this revenge!" said the Captain to the Canadian.

Ned bowed without replying. The combat had lasted a quarter of an hour. The monsters, vanquished and mutilated, left us at last, and disappeared under the waves. Captain Nemo, covered with blood, nearly exhausted, gazed upon the sea that had swallowed up one of his companions, and great tears gathered in his eyes.

RIGHT: 'Before my eyes was a horrible monster, worthy to figure in the legends of the marvellous', from *Twenty Thousand Leagues Under the Sea*, illustrated by Alphonse de Neuville and Édouard Riou, 1871.
© Courtesy of Wikimedia Commons

'IN THE ABYSS'
BY HG WELLS
(1907)

Herbert George Wells (1866–1946) is best known for his novels *The Time Machine* (1895) and *The War of the Worlds* (1898). He is often referred to as the 'father of science fiction' for his melding of speculative technologies with pertinent social commentary on class, colonialism and social Darwinism.

Wells was born in Bromley in Kent, England, and began his career as a teacher. He later completed a BSc in zoology and published the *Text-Book of Biology* in 1893 before going on to write novels and short stories. Wells never referred to his own works as 'science fiction' as the term wasn't in common use at the time; rather, he considered them to be 'scientific romances', a term that encompasses the combination of heroic narrative and scientific detail present in his work.

Wells' short story 'In the Abyss' was first published in 1896 in *Pearson's Magazine*. It features a fictional submersible, which precedes the first real one by more than 30 years – the *Bathysphere*, designed by naturalist William Beebe and engineer Otis Barton and used in the first ever deep-sea voyages. 'In the Abyss' was also later included in Wells' 1897 short-story collection *The Plattner Story and Others*.

LEFT: llustration by Warwick Goble, from HG Wells' short story 'In the Abyss', first published in *Pearson's Magazine*, August 1896. © Courtesy of Wikimedia Commons

'In the Abyss'

The lieutenant stood in front of the steel sphere and gnawed a piece of pine splinter. "What do you think of it, Steevens?" he said.

"It's an idea," said Steevens, in the tone of one who keeps an open mind.

"I believe it will smash—flat," said the lieutenant.

"He seems to have calculated it all out pretty well," said Steevens, still impartial.

"But think of the pressure," said the lieutenant. "At the surface of the water it's fourteen pounds to the inch, thirty feet down it's double that; sixty, treble; ninety, four times; nine hundred, forty times; five thousand, three hundred—that's a mile—it's two hundred and forty times fourteen pounds; that's—let's see—thirty hundredweight—a ton and a half, Steevens; a ton and a half to the square inch. And the ocean where he's going is five miles deep. That's seven and a half—"

"Sounds a lot," said Steevens, "but it's jolly thick steel."

The lieutenant made no answer, but resumed his pine splinter. The object of their conversation was a huge ball of steel, having an exterior diameter of perhaps nine feet. It looked like the shot for some Titanic piece of artillery. It was elaborately nested in a monstrous scaffolding built into the framework of the vessel, and the gigantic spars that were presently to sling it overboard gave the stern of the ship an appearance that had raised the curiosity of every decent sailor who had sighted it, from the Pool of London to the Tropic of Capricorn. In two places, one above the other, the steel gave place to a couple of circular windows of enormously thick glass, and one of these, set in steel frame of great solidity, was now partially unscrewed. Both the men had seen the interior of this globe for the first time that morning. It was elaborately padded with air cushions, with little studs sunk between bulging pillows to work the simple mechanism of the affair. Everything was elaborately padded, even the Myers apparatus which was to absorb carbonic acid and replace the oxygen inspired by its tenant, when he had crept in by the glass manhole, and had been screwed in. It was so elaborately padded that a man might have been fired from a gun in it with perfect safety. And it had need to be, for presently a man was to crawl in through that glass manhole, to be screwed up tightly, and to be flung overboard, and to sink down—down—down, for five miles, even as the lieutenant said. It had taken the strongest hold of his imagination; it made him a bore at mess; and he found Steevens the new arrival aboard, a godsend to talk to about it, over and over again.

"It's my opinion," said the lieutenant, "that, that glass will simply bend in and bulge and smash, under a pressure of that sort. Daubrée has made rocks run like water under big pressures—and you mark my words—"

"If the glass did break in," said Steevens, "what then?"

"The water would shoot in like a jet of iron. Have you ever felt a straight jet of high pressure water? It would hit as hard as a bullet. It would simply smash him and flatten him. It would tear down his throat, and into his lungs; it would blow in his ears—"

"What a detailed imagination you have!" protested Steevens, who saw things vividly.

"It's a simple statement of the inevitable," said the lieutenant.

"And the globe?"

"Would just give out a few little bubbles, and it would settle down comfortably against the Day of Judgment, among the oozes and the bottom clay— with poor Elstead spread over his own smashed cushions like butter over bread."

He repeated this sentence as though he liked it very much. "Like butter over bread," he said.

"Having a look at the jigger?" said a voice, and Elstead stood behind them, spick and span in white, with a cigarette between his teeth, and his eyes smiling out of the shadow of his ample hat-brim. "What's that about bread and butter, Weybridge? Grumbling as usual about the insufficient pay of naval officers? It won't be more than a day now before I start. We are to get the slings ready to-day. This clean sky and gentle swell is just the kind of thing for swinging off a dozen tons of lead and iron, isn't it?"

"It won't affect you much," said Weybridge.

"No. Seventy or eighty feet

ABOVE: Cover page for HG Wells' short story 'In the Abyss' published in *Amazing Stories*, illustrated by Frank Rudolph Paul, 1926.
© Courtesy of Wikimedia Commons

down, and I shall be there in a dozen seconds, there's not a particle moving, though the wind shriek itself hoarse up above, and the water lifts halfway to the clouds. No. Down there—" He moved to the side of the ship and the other two followed him. All three leant forward on their elbows and stared down into the yellow-green water.

"Peace," said Elstead, finishing his thought aloud.

"Are you dead certain that clockwork will act?" asked Weybridge presently.

"It has worked thirty-five times," said Elstead. "It's bound to work."

"But if it doesn't?"

"Why shouldn't it?"

"I wouldn't go down in that confounded thing," said Weybridge, "for twenty thousand pounds."

"Cheerful chap you are," said Elstead, and spat sociably at a bubble below.

"I don't understand yet how you mean to work the thing," said Steevens.

"In the first place, I'm screwed into the sphere," said Elstead, "and when I've turned the electric light off on three times to show I'm cheerful, I'm swung out over the stern by that crane, with all those big lead sinkers slung below me. The top lead weight has a roller carrying a hundred fathoms of strong cord rolled up, and that's all that joins the sinkers to the sphere, except the slings that will be cut when the affair is dropped. We use cord rather than wire rope because it's easier to cut and more buoyant— necessary points, as you will see.

"Through each of these lead weights you notice there is a hole, and an iron rod will be run through that and will project six feet on the lower side. If that rod is rammed up from below, it knocks up a lever and sets the clockwork in motion at the side of the cylinder on which the cord winds.

"Very well. The whole affair is lowered gently into the water, and the slings are cut. The sphere floats—, with the air in it, it's lighter than water—, but the lead weights go down straight and the cord runs out. When the cord is all paid out, the sphere will go down too, pulled down by the cord."

"But why the cord?" asked Steevens. "Why not fasten the weights directly to the sphere?"

"Because of the smash down below. The whole affair will go rushing down, mile after mile, at a headlong pace at last. It would be knocked to pieces on the bottom if it wasn't for that cord. But the weights will hit the bottom, and directly they do, the buoyancy of the sphere will come into play. It will go on sinking slower and slower; come to stop at last, and then begin to float upward again.

"That's where the clockwork comes in. Directly the weights smash against the sea bottom, the rod will be knocked through and will kick up the clockwork, and the cord will be rewound on the reel. I shall be lugged down to the sea bottom. There I shall stay for half an hour, with the electric light on, looking about me. Then the clockwork will release a spring knife, the cord will be cut, and up I shall rush again, like a soda-water bubble. The cord itself will help the flotation."

"And if you should chance to hit a ship?" said Weybridge.

"I should come up at such a pace, I should go clean through it," said Elstead, "like a cannon ball. You needn't worry about that."

"And suppose some nimble crustacean should wriggle into your clockwork—"

"It would be a pressing sort of invitation for me to stop," said

Elstead, turning his back on the water and staring at the sphere.

They had swung Elstead overboard by eleven o'clock. The day was serenely bright and calm, with the horizon lost in haze. The electric glare in the little upper compartment beamed cheerfully three times. Then they let him down slowly to the surface of the water, and a sailor in the stern chains hung ready to cut the tackle that held the lead weights and the sphere together. The globe, which had looked so large on deck, looked the smallest thing conceivable under the stern of the ship. It rolled a little, and its two dark windows, which floated uppermost, seemed like eyes turned up in round wonderment at the people who crowded the rail. A voice wondered how Elstead liked the rolling. "Are you ready?" sang out the commander. "Ay, ay, sir!" "Then let her go!"

The rope of the tackle tightened against the blade and was cut, and an eddy rolled over the globe in a grotesquely helpless fashion. Someone waved a handkerchief, someone else tried an ineffectual cheer, a middy was counting slowly, "Eight, nine, ten!" Another roll, then a jerk and a splash the thing righted itself.

It seemed to be stationary for a moment, to grow rapidly smaller, and then the water closed over it, and it became visible, enlarged by refraction and dimmer, below the surface. Before one could count three it had disappeared. There was a flicker of white light far down in the water, that diminished to a speck and vanished. Then there was nothing but a depth of water going down into blackness, through which a shark was swimming.

Then suddenly the screw of the cruiser began to rotate, the water was crickled, the shark disappeared in a wrinkled confusion, and a torrent of foam rushed across the crystalline clearness that had swallowed up Elstead. "What's the idea?" said one A.B. to another.

"We're going to lay off about a couple of miles, 'fear he should hit us when he comes up," said his mate.

The ship steamed slowly to her new position. Aboard her almost everyone who was unoccupied remained watching the breathing swell into which the sphere had sunk. For the next half-hour it is doubtful if a word was spoken that did not bear directly or indirectly on Elstead. The December sun was now high in the sky, and the heat very considerable.

"He'll be cold enough down there," said Weybridge. "They say that below a certain depth sea water's always just about freezing."

"Where'll he come up?" asked Steevens. "I've lost my bearings."

"That's the spot," said the commander, who prided himself on his omniscience. He extended a precise finger south-eastward. "And this, I reckon, is pretty nearly the moment," he said. "He's been thirty-five minutes."

"How long dose it take to reach the bottom of the ocean?" asked Steevens.

"For a depth of five miles, and reckoning—as we did—an acceleration of two feet per second, both ways, is just about three-quarters of a minute."

"Then he's overdue," said Weybridge.

"Pretty nearly," said the commander. "I suppose it takes a few minutes for that cord of his to wind in."

"I forgot that," said Weybridge, evidently relieved.

And then began the suspense. A minute slowly dragged itself out, and no sphere shot out of the water. Another followed, and nothing broke the low oily swell. The sailors explained to one another that little point about the winding-in of the cord. The rigging was dotted with expectant faces. "Come up, Elstead!" called one hairy-chested salt impatiently, and the others caught it up, and shouted as though they were waiting for the curtain of a theatre to rise.

The commander glanced irritably at them.

"Of course, if the acceleration's less than two," he said, "he'll be all the longer. We aren't absolutely certain that was the proper figure. I'm no slavish believer in calculations."

Steevens agreed concisely. No one on the quarter-deck spoke for a couple of minutes. Then Steevens' watchcase clicked.

When, twenty-one minutes after the sun reached the zenith, they were still waiting for the globe to reappear, and not a man aboard had dared to whisper that hope was dead. It was Weybridge who first gave expression to that realisation. He spoke while the sound of eight bells still hung in the air. "I always distrusted that window," he said quite suddenly to Steevens.

"Good God!" said Steevens; "you don't think—?"

"Well!" said Weybridge, and left the rest to his imagination.

"I'm no great believer in calculations myself," said the commander dubiously, "so that I'm not altogether hopeless yet." And at midnight the gunboat was steaming slowly in a spiral round the spot where the globe had sunk, and the white beam of the electric light fled and halted and swept discontentedly onward again over the waste of phosphorescent waters under the little stars.

"If his window hasn't burst and smashed him," said Weybridge, "then it's a cursed sight worse, for his clockwork has gone wrong, and he's alive now, five miles under our feet, down there in the cold and dark, anchored in that little bubble of his, where never a ray of light has shone or a human being lived, since the waters were gathered together. He's there without food, feeling hungry and thirsty and scared, wondering whether he'll starve or stifle. Which will it be? The Myers apparatus is running out, I suppose. How long do they last?"

"Good heavens!" he exclaimed; "What little things we are! What daring little devils! Down there, miles and miles of water—all water, and all this empty water about us and this sky. Gulfs!" He threw his hands out, and as he did so, a little white streak swept noiselessly up the sky, travelled more slowly, stopped, became a motionless dot, as though a new star had fallen up into the sky. Then it went sliding back again and lost itself amidst the reflections of the stars and the white haze of the sea's phosphorescence.

At the sight he stopped, arm extended and mouth open. He shut his mouth, opened it again, and waved his arms with an impatient gesture. Then he turned, shouted "El-stead ahoy!" to the first watch, and went at a run to Lindley and the search-light. "I saw him," he said "Starboard there! His light's on, and he's just shot out of the water. Bring the light round. We ought to see him drifting, when he lifts on the swell."

But they never picked up the explorer until dawn. Then they almost ran him down. The crane was swung out and a boat's crew hooked the chain to the sphere. When they had shipped the sphere,

138 'In the Abyss' by HG Wells

they unscrewed the manhole and peered into the darkness of the interior (for the electric light chamber was intended to illuminate the water about the sphere, and was shut off entirely from its general cavity).

The air was very hot within the cavity, and the indiarubber at the lip of the manhole was soft. There was no answer to their eager questions and no sound of movement within. Elstead seemed to be lying motionless, crumpled in the bottom of the globe. The ship's doctor crawled in and lifted him out to the men outside. For a moment or so they did not know whether Elstead was alive or dead. His face, in the yellow light of the ship's lamps, glistened with perspiration. They carried him down to his own cabin.

He was not dead, they found, but in a state of absolute nervous collapse, and besides cruelly bruised. For some days he had to lie perfectly still. It was a week before he could tell his experiences.

Almost his first words were that he was going down again. The sphere would have to be altered, he said, in order to allow him to throw off the cord if need be, and that was all. He had, had the most marvellous experience. "You thought I should find nothing but ooze," he said. "You laughed at my explorations, and I've discovered a new world!" He told his story in disconnected fragments, and chiefly from the wrong end, so that it is impossible to re-tell it in his words. But what follows is the narrative of his experience.

It began atrociously, he said. Before the cord ran out, the thing kept rolling over. He felt like a frog in a football. He could see nothing but the crane and the sky overhead, with an occasional glimpse of people on the ships rail. He couldn't tell a bit which way the thing would roll next. Suddenly he would find his footing going up, and try to step, and over he went rolling, head over heels, and just anyhow, on the padding. Any other shape would have been more comfortable, but no other shape was to be relied upon under the huge pressure of the nethermost abyss.

Suddenly the swaying ceased; the globe righted, and when he had picked himself up, he saw the water all about him greeny-blue, with an attenuated light filtering down from above, and a shoal of little floating things went rushing up past him, as it seemed to him, towards the light. And even as he looked, it grew darker and darker, until the water above was as dark as the midnight sky, albeit of greener shade, and the water below black. And little transparent things in the water developed a faint glint of luminosity, and shot past him in faint greenish streaks.

And the feeling of falling! It was just like the start of a lift, he said, only it kept on. One has to imagine what that means, that keeping on. It was then of all times that Elstead repented of his adventure. He saw the chances against him in an altogether new light. He thought of the big cuttle-fish people knew to exist in the middle waters, the kind of things they find half digested in whales at times, or floating dead and rotten and half eaten by fish. Suppose one caught hold and wouldn't let go. And had the clockwork really been sufficiently tested? But whether he wanted to go on or go back mattered not the slightest now.

In fifty seconds everything was as black as night outside, except where the beam from his light struck through the waters,

ABOVE: Portrait of HG Wells *c.* 1918.
© Courtesy of Wikimedia Commons

and picked out every now and then some fish or scrap of sinking matter. They flashed by too fast for him to see what they were. Once he thinks he passed a shark. And then the sphere began to get hot by friction against the water. They had underestimated this, it seems.

The first thing he noticed was that he was perspiring, and then he heard a hissing growing louder under his feet, and saw a lot of little bubbles—very little bubbles they were—rushing upward like a fan through the water outside. Steam! He felt the window, and it was hot. He turned on the minute glow-lamp that lit his own cavity, looked at the padded watch by the studs, and saw he had been travelling now for two minutes. It came into his head that the window would crack through the conflict of temperatures, for he knew the bottom water is very near freezing.

Then suddenly the floor of the sphere seemed to press against his feet, the rush of bubbles outside grew slower and slower, and the hissing diminished. The sphere rolled a little. The window had not cracked, nothing had given, and he knew that the dangers of sinking, at any rate, were over.

In another minute or so he would be on the floor of the abyss. He thought, he said, of Steevens and Weybridge and the rest of them five miles overhead, higher to him than the highest clouds that ever floated over land are to us, steaming slowly and staring down and wondering what had happened to him.

He peered out of the window. There were no more bubbles now, and the hissing had stopped. Outside there was a heavy blackness—as black as black velvet—except where the electric light pierced the empty water and showed the colour of it—a yellow-green. Then three things like shapes of fire swam into sight, following each other through the water. Whether they were little and near or big and far off he could not tell.

Each was outlined in a bluish light almost as bright as the lights of a fishing smack, a light which seemed to be smoking greatly, and all along the sides of them were specks of this, like the lighter portholes of a ship. Their phosphorescence seemed to go out as they came into the radiance of his lamp, and he saw then that they were little fish of some strange sort, with huge heads, vast eyes, and dwindling bodies and tails. Their eyes were turned towards him, and he judged they were following him down. He supposed they were attracted by his glare.

Presently others of the same sort joined them. As he went on down, he noticed that the water became of a pallid colour, and that little specks twinkled in his ray like motes in a sunbeam. This was probably due to the clouds of ooze and mud that the impact of his leaden sinkers had disturbed.

By the time he was drawn down to the lead weights he was in a dense fog of white that his electric light failed altogether to pierce for more than a few yards, and many minutes elapsed before the hanging sheets of sediment subsided to any extent. Then, lit by his light and by the transient phosphorescence of a distant shoal of fishes, he was able to see under the huge blackness of the super-incumbent water an undulting expanse of greyish-white ooze, broken here and there by tangled thickets of a growth of sea lilies, waving hungry tentacles in the air.

Farther away were the graceful, translucent outlines of a group of gigantic sponges. About this floor there were scattered a number of bristling flattish tufts of rich purple and black, which he decided must be some sort of sea-urchin, and small, large-eyed or blind things having a curious resemblance, some to woodlice, and others to lobsters, crawled sluggishly across the track of the light and vanished into the obscurity again, leaving furrowed trails behind them.

Then suddenly the hovering swarm of little fishes veered about and came towards him as a flight of starlings might do. They passed over him like a phosphorescent snow, and then he saw behind them some larger creature advancing towards the sphere.

At first he could see it only dimly, a faintly moving figure remotely suggestive of a walking man, and then it came into the spray of light that the lamp shot out. As the glare struck it, it shut its eyes, dazzled. He stared in rigid astonishment.

It was a strange vertebrated animal. Its dark purple head was dimly suggestive of a chameleon, but it had such a high forehead and such a braincase as no reptile ever displayed before; the vertical pitch of its face gave it a most extraordinary resemblance to a human being.

Two large and protruding eyes projected from sockets in chameleon fashion, and it had a broad reptilian mouth with horny lips beneath its little nostrils. In the position of the ears were two huge gill-covers, and out of these floated a branching tree of coralline filaments, almost like the tree-like gills that very young rays and sharks possess.

But the humanity of the face was not the most extraordinary thing about the creature. It was a biped; its almost globular body was poised on a tripod of two frog-like legs and a long thick tail, and its fore limbs, which grotesquely caricatured the human hand, much as a frog's do, carried a long shaft of bone, tipped with copper. The colour of the creature was variegated; its head, hands and legs were purple; but its skin, which hung loosely upon it, even as clothes might do, was a phosphorescent grey. And it stood there blinded by the light.

At last this unknown creature of the abyss blinked its eyes open, and shading them with its disengaged hand, opened its mouth and gave vent to a shouting noise, articulate almost as speech might be, that penetrated even the steel case and padded jacket of the sphere. How a shouting may be accomplished without lungs Elstead dose not profess to explain. It then moved sideways out of the glare into the mystery of shadow that bordered it on either side, and Elstead felt rather than saw that it was coming towards him. Fancying the light had attracted it, he turned the switch that cut off the current. In another moment something soft dabbed upon the steel, and the globe swayed.

Then the shouting was repeated, and it seemed to him that a distant echo answered it. The dabbing recurred, and the whole globe swayed and ground against the spindle over which the wire was rolled. He stood in the blackness and peered out into the everlasting night of the abyss. And presently he saw, very faint and remote, other phosphorescent quasi-human forms hurrying towards him.

Hardly knowing what he did, he felt about in his swaying prison for the stud of the exterior electric light, and came by accident against his own small glow-lamp in its padded recess. The sphere twisted, and then threw him down; he heard shouts like shouts of surprise, and when he rose to his feet, he saw two pairs of stalked eyes peering into the lower window and reflecting his light.

In another moment hands were dabbing vigorously at his steel casing, and there was a sound, horrible enough in his position, of the metal protection of the clockwork being vigorously hammered. That indeed sent his heart into his mouth, for if these strange creatures succeeded in stopping that, his release would never occur. Scarcely had he thought as much when he felt the sphere sway violently, and the floor of it press hard against his feet. He turned off the small glow-lamp that lit the interior, and sent the ray of the large light in the separate compartment, out into the water. The sea-floor and the man-like creatures had disappeared, and a couple of fish chasing each other dropped suddenly by the window.

He thought at once that these strange denizens of the deep sea had broke the rope, and that he had escaped. He drove up faster and faster, and then stopped with a jerk that sent him flying against the padded roof of his prison. For half a minute perhaps, he was too astonished to think.

Then he felt that the sphere was spinning slowly, and rocking, and it seemed to him that it was also being drawn through the water. By crouching close to the window, he managed to make his weight effective and roll that part of the sphere downward, but he could see nothing save the pale ray of his light striking down ineffectively into the darkness. It occurred to him that he would see more if he turned the lamp off, and allowed his eyes to grow accustomed to the profound obscurity.

In this he was wise. After some minutes the velvety blackness became a translucent blackness, and then, far away, and as faint as zodiacal light of an English summer evening, he saw shapes moving below. He judged these creatures had detached his cable, and were towing him along the sea bottom.

And then he saw something faint and remote across the undulations of the submarine plain, a broad horizon of pale luminosity that extended this way and that way as far as the range of his little window permitted him to see. To this he was being towed, as a balloon might be towed by men out of the open country into a town. He approached it very slowly, and very slowly the dim irradiation was gathered together into more definite shapes.

It was nearly five o'clock before he came over this luminous area, and by that time he could make out an arrangement suggestive of streets and houses grouped about a vast roofless erection that was grotesquely suggestive of a ruined abbey. It was spread out like a map below him. The houses were all roofless enclosures of walls, and their substance being, as he afterwards saw, of phosphorescent bones, gave the place an appearance as if it were built of drowned moonshine.

Among the inner caves of the place waving trees of crinoid stretched their tentacles, and tall, slender, glassy sponges shot like shining minarets and lilies of filmy light out of the general glow of the city. In the open spaces of the place he could see a stirring movement as of crowds of people, but he was too many fathoms above them to distinguish the individuals in those crowds.

Then slowly they pulled him down, and as they did so, the details of the place crept slowly upon his apprehension. He saw that the courses of the cloudy buildings were marked out with beaded lines of round objects, and then he perceived that at several points below him, in broad open spaces, were forms like the encrusted shapes of ships.

Slowly and surely he was drawn down, and the forms below him became brighter, clearer, more distinct. He was being pulled down, he perceived, towards the large building in the centre of the town, and he could catch a glimpse ever and again of the multitudinous forms that were lugging at his cord. He was astonished to see that the rigging of one of the ships, which formed such a prominent feature of the place, was crowded with a host of gesticulating figures regarding him, and then the walls of the great building rose about him silently, and hid the city from his eyes.

And such walls they were, of water-logged wood, and twisted wire-rope, and iron spars, and copper, and the bones and skulls of dead men. The skulls ran in zigzag lines and spirals and fantastic curves over the building; and in and out of their eye-sockets, and over the whole surface of the place, lurked and played a multitude of silvery little fishes.

Suddenly his ears were filled with a low shouting and a noise like the violent blowing of horns, and this gave place to a fantastic chant. Down the sphere sank, past the huge pointed windows, through which he saw vaguely a great number of these strange, ghostlike people regarding him, and at last he came to rest, as it seemed, on a kind of altar that stood in the centre of the place.

And now he was at such a level that he could see these strange people of the abyss plainly once more. To his astonishment, he perceived that they were prostrating themselves before him, all save one, dressed as it seemed in a robe of placoid scales, and crowned with a luminous diadem, who stood with his reptilian mouth opening and shutting, as though he led the chanting of the worshippers.

A curious impulse made Elstead turn on his small glow-lamp again, so that he became visible to these creatures of the abyss, albeit the glare made them disappear forthwith into night. At this sudden sight of him, the chanting gave place to a tumult of exultant shouts; and Elstead, being anxious to watch them, turned his light off again, and vanished from before their eyes. But for a time he was too blind to make out what they were doing, and when at last he could distinguish them, they were kneeling again. And thus they continued worshipping him, without rest or intermission, for a space of three hours.

Most circumstantial was

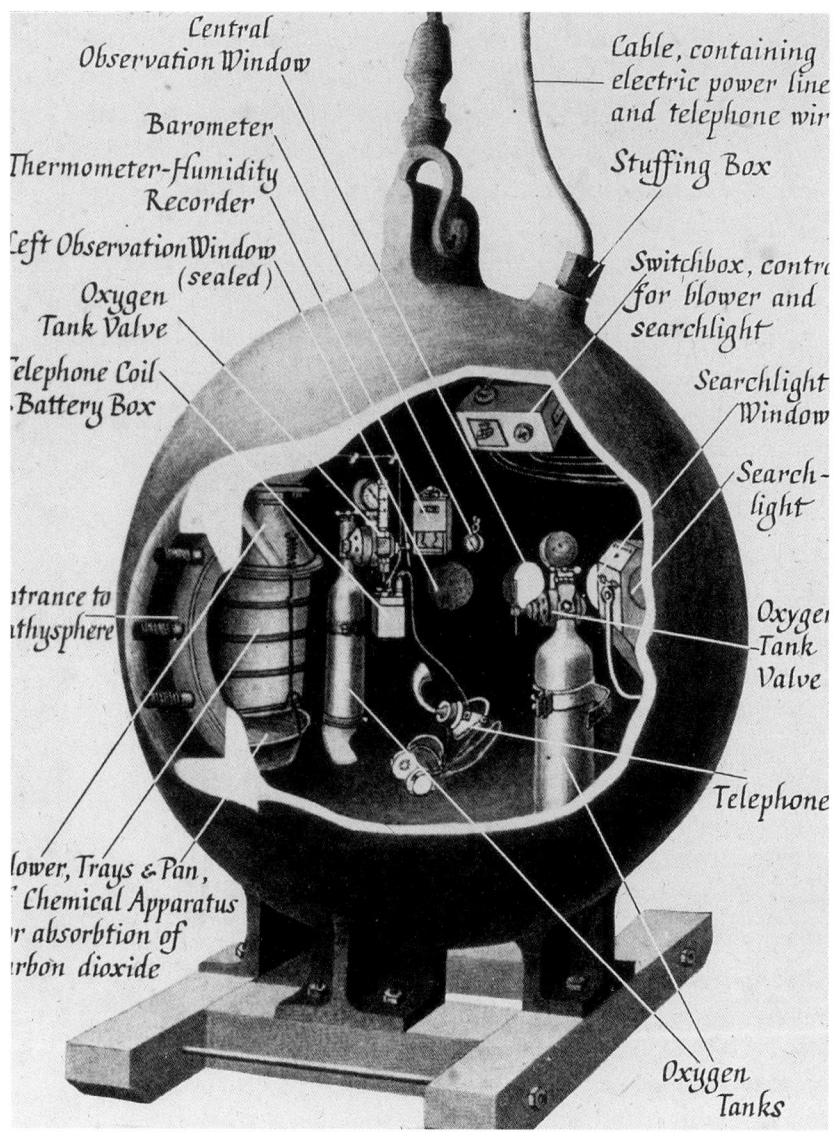

ABOVE: 'All instruments aboard' showing the parts of William Beebe's *Bathysphere*, illustrated by Charles E Riddiford, 1934.
© Courtesy of ETH Library Zürich, Image Archive / Dia_247-F-00744

Elstead's account of this astounding city and its people, these people of perpetual night, who have never seen sun or moon or stars, green vegetation, nor any living, air-breathing creatures, who know nothing of fire, nor any light but the phosphorescent light of living things.

Startling as is his story, it is yet more startling to find that scientific men, of such eminence as Adams and Jenkins, find nothing incredible in it. They tell me they see no reason why intelligent, water-breathing, vertebrated creatures, inured to a low temperature and enormous pressure, and of such a heavy structure, that neither alive nor dead would they float, might not live upon the bottom of the deep sea, and quite unsuspected by us, descendants like ourselves of the great Theriomorpha of the New Red Sandstone age.

We should be known to them however, as strange, meteoric creatures, wont to fall catastrophically dead out of the mysterious blackness of their watery sky. And not only we ourselves, but our ships, our metals, our appliances, would come raining down out of the night. Sometimes sinking things would smite down and crush them, as if it were the judgment of some unseen power above, and sometimes would come things of utmost rarity or utility, or shapes of inspiring suggestion. One can understand, perhaps, something of their behaviour at the descent of a living man, if one thinks what a barbaric people might do, to whom an enhaloed, shining creature came suddenly out of the sky.

At one time or another Elstead probably told the officers of the *Ptarmigan* every detail of his strange twelve hours in the abyss. That he also intended to write them down is certain, but he never did, and so unhappily we have to piece together the discrepant fragments of his story from the reminiscences of Commander Simmons, Weybridge, Steevens, Lindley, and the others.

We see the thing darkly in fragmentary glimpses—the huge ghostly building, the bowing, chanting people, with their dark chameleon-like heads and faintly luminous clothing, and Elstead, with his light turned on again, vainly trying to convey to their minds that the cord by which the sphere was held was to be severed. Minute after minute slipped away, and Elstead, looking at his watch, was horrified to find that he had oxygen only for four hours more. But the chant in his honour kept on as remorselessly as if it was the marching song of his approaching death.

The manner of his release he does not understand, but to judge by the end of cord that hung from the sphere, it had been cut through by rubbing against the edge of the altar. Abruptly the sphere rolled over, and he swept up, out of their world, as an ethereal creature clothed in a vacuum would sweep through our own atmosphere back to its native ether again. He must have torn out of their sight as a hydrogen bubble hastens upwards from our air. A strange ascension it must have seemed to them.

The sphere rushed up with even greater velocity than, when weighted with the lead sinkers, it had rushed down. It became exceedingly hot. It drove up with the windows uppermost, and he remembers the torrent of bubbles frothing against the glass. Every moment he expected this to fly. Then suddenly something like a huge wheel seemed to be released in his head, the padded compartment began spinning about him, and he fainted. His next recollection was of his cabin, and of the doctor's voice.

But that is the substance of the extraordinary story that Elstead related in fragments to the officers of the *Ptarmigan*. He promised to write it all down at a later date. His mind was chiefly occupied with the improvement of his apparatus, which was effected at Rio.

It remains only to tell that on February 2, 1896, he made his second descent into the ocean abyss, with the improvements his first experience suggested. What happened we shall probably never know. He never returned. The *Ptarmigan* beat about over the point of his submersion, seeking him in vain for thirteen days. Then she returned to Rio, and the news was telegraphed to his friends. So the matter remains for the present. But it is hardly probable that no further attempt will be made to verify his strange story of these hitherto unsuspected cities of the deep sea.

'THE TERROR OF THE SEA CAVES'

BY SIR CHARLES GD ROBERTS
(1907)

Charles GD Roberts (1860–1943) is best known for his nature poetry and his prominence in the Canadian Romantic movement. His poems draw upon the forms and nature themes of English Romanticism while shifting the subject to the forests, lakes and coasts of New Brunswick and Nova Scotia. The publication of his first poetry collection, *Orion, and Other Poems*, in 1880 helped to establish Roberts as a member of the nationalist literary group known as the Confederation Poets.

In addition to his poetry collections, Roberts published multiple novels, guidebooks and histories of Canada, and a collection of stories about animals called *Earth's Enigmas* (1896). These 'animal stories' are notable for their lyricism and non-human points of view, and were collected in *The Haunters of the Silences: A Book on Animal Life* (1907) and *Eyes of the Wilderness* (1933). Roberts served in the First World War, first in the British Army then the Canadian Army. He was awarded a knighthood in 1935.

'The Terror of the Sea Caves' was originally published in the January 1907 issue of *Everybody's Magazine* and then included in Roberts' collection of animal stories *The Haunters of the Silences*. It is particularly remarkable for the third part, which is narrated from the perspective of a giant squid, lending a sympathetic tentacle to an otherwise stock monster of the oceanic horror stories of the time.

LEFT: 'There was a great variety of tall, noble trees, loaded with marine fruit, such as lobsters, crabs, oysters', from *The Surprising Adventures of Baron Munchausen*, illustrated by Joseph Benwell Clark, 1895.
© Courtesy of Getty Images

'The Terror of the Sea Caves'

I.

IT was in Singapore that big Jan Laurvik, the diver, heard about the lost pearls.

As he was passing the head of a mean-looking alley near the waterside, late one sweltering afternoon, he was halted by a sudden uproar of cries and curses. The noise came from a courtyard about twenty paces up the alley. It was a fight evidently, and Jan's blood responded with a sympathetic thrill. But the curses which he caught were all in Malay or Chinese, and he curbed his natural desire to rush in and help somebody. Though he knew both languages very well, he knew that he did not know, and never could know, the people who spoke those languages. Interference on the part of a stranger might be resented by both parties to the quarrel. He shrugged his great shoulders and walked on reluctantly.

Hardly three steps had he taken, however, when above the shrill cries a great voice shouted.

"Take that, you ——" it began, in English. And at that it ended, with a kind of choking.

Jan Laurvik wheeled round in a flash and ran curiously for the door of the courtyard, which stood half open. He was a Norwegian, but English was as a native tongue to him; and amid the jumble of races in the East he counted all of European speech his brothers. An Englishman was being killed in there, The quarrel was clearly his.

Six feet two in height, swift, and of huge strength, with yellow hair, so light as to be almost white, waving thickly over a face that was sunburnt to a high red, his blue eyes flaming with the delight of battle, Jan burst in upon the mob of fighters. Several bodies lay on the floor.

One dark-faced, low-browed fellow, a Lascar apparently, with his back to the wall and a bloody kreese in his hand, was putting up a savage fight against five or six assailants, who seemed to be Chinamen and Malays. The body of the Englishman whose voice Jan had heard lay in an ugly heap against the wall, its head far back and almost severed.

Jan's practised eye took in everything at a glance. The heavy stick he carried was, for a mêlée like this, a better weapon than knife or gun. With a great bellowing you he sprang upon the knot of fighters.

The result was almost instantaneous. The two newest rascals went down at his first two strokes. At the sound of that huge roar of his all had turned their eyes; and the man at bay, seizing his opportunity, had cut down two more of his foes with lightning slashes of his blade. The remaining two, scattering and ducking, had leaped for the door like rabbits. Jan wheeled, and sprang after them. But they were too quick for him. As he reached the head of the alley they darted into a narrow doorway across the street which led into a regular warren of low structures. Knowing it would be madness to follow, Jan turned back to the courtyard, curious to find out what it had all been about.

The silence was now startling. As he entered, there was no sound but the painful breathing of the Lascar, whom he found sitting with his back against the wall, close beside the body of the Englishman. He was desperately slashed. His eyes were half closed; and Jan saw that there was little chance of his recovery. Besides that of the Englishman, there were six bodies lying on the floor, all apparently quite lifeless. Jan saw that the place was a kind of drinking proprietor, a brutal-looking Chinaman, lay dead beside his jugs and bottles. Jan reached for jug of familiar appearance, poured out a cup of arrack and held it to the lips of the dying Lascar. At the first gulp of the potent spirit his eyes opened again. He swallowed it all eagerly, then straightened himself up, held out his hand in European fashion to Jan, and thanked him in Malayan.

"Who's that?" enquired Jan in the same tongue, pointing to the dead white man.

Grief and rage convulsed the fierce face of the wounded Lascar.

"He was my friend," he answered. "The sons of filthy mothers, they killed him!"

"Too bad!" said Jan sympathetically. "But you gave a pretty good account of yourselves, you two. I like a man that can fight like you were fighting when I came in. What can I do for you?"

"I'm dead, pretty soon now!" said the fellow indifferently.

And from the blood that was soaking down his shirt and spreading on the floor about him, Jan saw that the words were true. Anxious, however, to do something to show his goodwill, he pulled out his big red handkerchief, and knelt to bandage a gaping slash straight across the man's left forearm, from which the bright arterial blood was jumping hotly. As he bent, the fellow's eyes lifted and looked over his shoulder.

"Look out!" he screamed. Before the words were fairly out of his mouth Jan had thrown himself violently to one side and sprung to his feet. He was just in time. The knife of one of the Chinamen whom he had supposed to be dead was sticking in the wall beside the Lascar's arm.

Jan stared at the bodies — all, apparently, lifeless.

"That's the one did it," cried the Lascar excitedly, pointing to one whom Jan had struck on the head with his stick. "Put your knife into the son of a dog!"

But that was not the big Norseman's way. He wanted to assure himself. He went and bent over the limp-looking, sprawling shape, to examine it. As he did so the slant eyes

ABOVE: Inside cover illustration from *The Haunters of the Silences: A Book of Animal Life,* illustrated by Charles Livingston Bull, 1907.
© Courtesy of Internet Archive

'The Terror of the Sea Caves' by Sir Charles GD Roberts 149 opened upon his with a flash of such maniacal hate that he started back. He was just in time to save his eyes, for the Chinaman had clutched at them like lightning with his long nails.

Startled and furious at this novel attack, Jan reached for his knife. But before he could get his hand on it the Chinaman had leaped into the air like a wild-cat wound arms and legs about his body, and was struggling like a mad beast to set teeth into his throat. The attack was so miraculously swift, so disconcerting in its beast-like ferocity, that Jan felt a strange qualm that was almost akin to panic. Then a black rage swelled his muscles; and tearing the creature from him he dashed him down upon the floor, on the back of his neck, with a violence which left no need of pursuing the question further. Not till he had examined each of the bodies carefully, and tried them with his knife, did he turn again to the wounded Lascar leaning against the wall.

"Thank you, my friend!" he said simply.

"You're a good fighting man. You're — like him," answered the Lascar feebly, nodding toward the dead Englishman.

"Give me more arrack. I will tell you something. Hurry, for I go soon."

Jan brought him the liquor, and he gulped it. Then from a pouch within his knotted silk waistband he hurriedly produced a bit of paper which he unfolded with trembling fingers. Jan saw that it was a rough map sketched with India ink and marked with Malayan characters. The Lascar peered about him with fierce eyes already growing dim.

"Are you sure they are all gone?" he demanded.

"Certain!" answered Jan, highly interested.

"They'll try their best to kill you," went on the dying man. "Don't let them. If you let them get the pearls, I'll come back and haunt you."

"I won't let them kill me, and I won't let them get the pearls, if that's what it is that's made all the trouble. Don't worry about that," responded Jan confidently, reaching out his great hand for the paper, which was evidently so precious that men were giving up their lives for it.

The man handed it over with a groping gesture, though his savage black eyes were wide open.

"That'll show you where the wreck of the junk lies, in seven or eight fathoms of water, close inshore. The pearls are in the deck-house. He kept them. The steamer was on a reef, going to pieces, and we came up just as the boats, were putting off. We sunk them all, and got the pearls. And next night, in a storm, the junk was carried on to the rocks by a current we didn't know about. Only five of us got ashore — for the sharks were around, and the "killers", that night. Him and me, we were the only ones knew enough to make that map."

Here the dying pirate — for such he had declared himself — sank forward with his face upon his knees. But with a mighty effort he sat up again and fixed Jan Laurvik with terrible eyes.

"Don't let the sons of a dog get them, or I will come back and choke you in your sleep," he gasped, suddenly pointing a lean finger straight at the Norseman's face. Then his black eyes opened wide, a strange red light blazed up in them for an instant and faded. With a sigh he toppled over, dead, his head resting on the dead Englishman's feet.

150 'The Terror of the Sea Caves' by Sir Charles GD Roberts

II.

Jan Laurvik looked down upon the slack form with a sort of grim indulgence. "He was game, and he loved his comrade, though he was but a bloodthirsty pirate!" he muttered to himself.

With the paper; folded small and hidden in his great palm, he glanced again from the door to see if any of the routed were coming back. Satisfied on this point, he once more investigated the dead bodies on the floor, to assure himself that all were as dead as they appeared. Then he set himself to examine the precious paper, which held out to his imagination all sorts of fascinating possibilities. He knew that the swift boats carrying the proceeds of the pearl fisheries were always eagerly watched by the piratical junks infesting those waters, but carried an armament which secured them from all interference. In case of wreck however, the pirates' opportunity would come. Jan knew that the story he had just heard was no improbable one.

The map proved to be rough, but very intelligible. It indicated a stretch of the eastern coast of Java, which Jan recognized; but the spot where the junk had gone down was one to which passing ships always gave a wide berth. It was a place of treacherous anchorage, of abrupt, forbidding, uninhabited shore, and of violent currents that shifted erratically. So much the better, thought Jan, for his investigations, if only the pirate junk should prove to have been considerate enough to sink in water not too deep for a diver to work in. There would be so much the less danger of interruption.

Jan was on the point of hurrying away from the gruesome scene, which might at any moment become a scene of excitement and annoying investigation, when a new idea flashed into his mind. It was over this precious paper that all the trouble had been. The scoundrels who had fled would undoubtedly return as soon as they dared and would search for it. Finding it gone they would conclude that he had it and they would be hot on his trail. He had no fancy for the sleepless vigilance that this would entail upon him. He had no fancy for the heavy armed expedition which it would force him to organize for the pearl hunt. He saw his airy palaces toppling ignominiously to earth. He saw that all he was likely to get was a slit throat.

As he glanced about him for a way out of his dilemma his eyes fell on a bottle of India ink containing the fine-tipped brush with which these Orientals did their writing. His resourcefulness awoke to this chance. The moments were becoming very pearls themselves for preciousness but seizing the brush, he made a workable copy of the map on the back of a letter which he had in his pocket. Then he made a minute and very careful correction in the original, in such a manner as to indicate that the position of the wreck was in a deep fiord some fifty miles east of where it actually was. This done to his critical satisfaction, he returned the map to its hiding-place in the dead pirate's belt, and made all haste away. Not till he was back in the European quarter did he feel himself secure. Once among his fellow whites, where he was a man of known standing and reputed to be the best diver in the Archipelago, he knew that he would run no

ABOVE: Illustration from page 302 of *The Haunters of the Silences: A Book of Animal Life*, illustrated by Charles Livingston Bull, 1907.
© Courtesy of Internet Archive

risk of being connected with a drinking brawl of Lascars and pirates. As for the dead Englishman, he knew the odds were that the Singapore police would know all about him.

Jan Laurvik had a little capital. But he needed a trusty partner with more. To his experienced wits his other needs were clear. There would have to be a very seaworthy little steamer, powerfully engined for service on that stormy coast, and armed to defend herself against prowling pirate junks. This small and fit craft would have to be manned by a crew equally fit, and at the same time as small as possible, for the reason that in a venture of this sort everyone concerned would of necessity come in for a share of the winnings. Moreover, the fewer there were to know, the fewer the chances of the secret leaking out; and Jan was even more in dread of the Dutch Government getting wind of it than he was of the pirates picking up his trail.

Up to a certain point, he had no difficulty in verifying the dead pirate's story. He had heard of the wreck of the Dutch steamer *Viecht* on a reef off the Celebes, and of the massacre of all the crew and passengers, except one small boatload, by pirates. This had happened about eight months ago. Discreet enquiry developed the fact that the *Viecht* had carried about $300,000 worth of pearls. The evidence was sufficiently convincing, and the prize was sufficiently alluring to make it worth his while to risk the adventure.

It was with a certain amount of Northern deliberation that Jan Laurvik thought these points all out, and made up his mind what to do. Then he acted promptly. First he cabled to Calcutta, to one Captain Jerry Parsons, to join him in Singapore without fail by the very next steamer. Then he set himself unobtrusively to the task of finding the craft he wanted and looking up the equipment for her.

Captain Jerry Parsons was a New Englander, from Portland, Maine. He had been whaler, gold-hunter, filibuster, copra-trader, general-in-chief to a small Central American republic, and sheep-farmer in the Australian bush. At present he was conducting a more or less regular trade in precious stones among the lesser Indian potentates. He loved gain much, but he loved adventure more.

When he received the cable from his good friend Jan Laurvik, he knew that both were beckoning to him. With light-hearted zest he betook himself to the steamship offices, found a P. and O. boat sailing on the morrow, and booked his passage. Throughout the journey he amused himself with

152 'The Terror of the Sea Caves' by Sir Charles GD Roberts

trying to guess what Jan Laurvik was after, and, as it happened, almost the only thing he failed to think of was pearls.

When Captain Jerry reached Singapore Jan Laurvik told him the story of the dead pirate's map.

"Let's see the map!" said he, chewing hard on the butt of his unlighted Manila.

Jan passed his copy over. The New Englander inspected it carefully, in silence, for several minutes.

"'Tain't much of a map!" said he at length disparagingly. "You think the varmint was straight?"

"In his way, yes," answered Jan with conviction. "He had it in him to be straight in his way to a friend, which wouldn't hinder him cuttin' the throats of a thousand chaps he didn't take an interest in."

"When shall we start?" asked Captain Jerry. Now that his mind was quite made up he took out his matchbox and carefully lighted his cheroot.

The big Norseman's face lighted up with pleasure, and he reached out his hand. The grip was all, in the way of a bargain, that was needed between them.

"Why, tomorrow night!" he answered.

"Well," said the New Englander, "I'll draw some cash in the morning."

The boat which Jan had hired was a fast and sturdy sea-going tug, serviceable, but not designed for comfort. Jan had retained her engineer, a shrewd and close-mouthed Scotsman. Her sailing-master would be Captain Jerry. For crew he had chosen a wiry little Welshman and two lank leather-skinned Yankees. To these four, for whose honesty and loyalty he trusted to his own insight as a reader of men, he explained, partially, the nature of the undertaking, and agreed to give them, over and above their wages, a substantial percentage of whatever treasure he might succeed in recovering. He had made his selection wisely, and every man of the four laid hold of the opportunity with ardour.

The tug was swift enough to elude any of the junks infesting those waters, but the danger was that she might be taken by surprise at her anchorage while Laurvik was under water. He fitted her, therefore, with a Maxim gun on the roof of the deck-house, and armed the crew with repeating Winchesters.

Thus equipped, he felt ready for any perils that might confront him above the surface of the water. As to what might lurk below he felt somewhat less confident, as these he should have to face alone, and he remembered the ominous warning of his pirate friend, about the sharks and the "killers". For sharks Jan Laurvik had comparatively small concern; but for the "killers", those swift and implacable little whales who fear no living thing, he entertained the highest respect.

On the evening of the day after Captain Jerry's arrival, the tug *Sarawak* steamed quietly out of the harbour. As this was a customary thing for her to do, it excited no particular comment among the frequenters of the waterside. By the pirates' spies, who abounded in the city, it was not considered an event worth noting.

The journey, across the Straits, and down the treacherous Javan Sea, was so prosperous that Jan Laurvik, his blood steeped in Norse superstition, began to feel uneasy. The sea was like a millpond all the way, and they were sighted by no one likely to interfere or ask questions. Jan distrusted Fortune when she seemed to smile too blandly. But Captain Jerry comforted him with the assurance that there'd be trouble enough ahead; and strangely enough, this singular variety of comfort quite relieved Jan's depression.

The unusual calm made it easy to hold close inshore, when they reached that portion of the coast where they must keep watch for the landmarks indicated on the pirate's map. Every reef and surface-ledge boiled ceaselessly in the smooth swell, and by that clear green sea they were saved the trouble of tedious soundings. When they came exactly abreast of a low headland which they had been watching for some time, it suddenly opened out into the semblance of a two-humped camel crouching sidewise to the sea, exactly as it was represented in Jan's map. Just beyond was a narrow bay, and across the middle of its mouth, with a dangerous passage on either side, stretched the reef on which the pirate junk had gone down. At this hour of low water the reef was showing its teeth and snarling with surf. At high tide it would be hidden, and a perfect snare of ships. According to the map, the wreck lay in some eight fathoms of water, midway of the outer crescent of the reef. Behind the reef, where the latter might serve them as a partial shelter from the sweep of the seas if a north-easter should blow up, they found tolerable anchorage for the tug. For the preliminary soundings, and for the diving operations, of course, Jan planned to use the launch. And, in order to take utmost advantage of the phenomenal calm, which seemed determined to smooth away every obstacle for the adventurers, Jan got instantly to work. Within a half-hour of the *Sarawak*'s anchoring he had the launch outside the reef with all his diving apparatus aboard, with Captain Jerry to manage the air-pump, and the Scottish engineer to run the motor.

III.

Along the outer face of the reef, at a depth varying from eight to twelve fathoms, ran an irregular rocky shelf which dipped gradually seaward for several hundred yards, then dropped sheer to the ocean depths. In the warm water along this shelf swarmed a teeming life, of gay-coloured gigantic weeds and of strange fish that outdid the brightest weeds in brilliancy and unexpectedness of hue. Where the tropic sunlight filtered dimly down through the beryl tide it sank into a marvellous garden whose flowers, for the most part, were living and moving forms some monstrous, many terrifying, and almost all as grotesque in shape as they were radiant in colour. But in that insufficient, glimmering light, which was rather, to a human eye a vaguely translucent, greenish darkness, these colours were almost blotted out. It took eyes adapted to the depth and gloom to differentiate them clearly.

In the great deeps, also, beyond the edge of the shelf, thronged life in swimming, crawling, or moveless forms, of every imagined and many unimagined shapes, from creatures so tiny that a whole colony could dwell at ease in the eye of a cambric needle, to the titanic squid, or cuttlefish, with oval body fifty feet in length and arms like writhing constrictors reaching twenty or thirty feet further. It was a life of noiseless but terrific activity, of unrelenting and incessant death, in a darkness streaked fitfully with phosphorescent gleams from the bodies of the darting, writhing, or pouncing creatures that slew and were slain in the stupendous silence.

Down to these dwellers in the profound had come some mysterious message or exciting influence, no man knows what, from the prolonged calm on the surface. It affected individuals among various species, in such a way that they moved upward, into a twilight where they were aliens and intruders. Among those so stung with unrest were several of the gigantic, pallid cuttles. Far offshore, one of these monsters came up and sprawled upon the surface in the unfriendly sun, his dreadful arms curling and uncurling like snakes, till a great sperm-whale, of scarcely more than his own size, came by and fell upon him ravenously, and devoured him.

Another of the restless monsters, however, kept his restlessness within the bounds of discretion. Slowly rising, a vast and spectral horror as he came up into the green light, he reached the rim of the ledge. The growing light had already made him uneasy, and he wanted no more of it. Here on the ledge, where food, though novel in character, was unlimited in supply, was variety enough to content him. Gorging himself as he went with everything that swam within reach of his darting tentacles, he moved over the rocky floor till he came to the wreck of the junk.

To his huge unwinking eyes of crystal black, which caught every tiniest ray of light in their smooth, appalling deeps, the wreck looked strange enough to attract his attention at once. It was quite unlike any rock-form which he had ever seen. Rather cautiously he advanced a giant tentacle to investigate it. But at the touch of the unfamiliar and alien substance the tentacle recoiled in aversion. The pale monster backed away. But the wreck made no attempt to pounce upon him. It seemed to have no fight in it. Possibly, on closer investigation, it might prove to be good to eat; and he was hungry. In fact, he was always hungry, for the irresistible corrosives in his great stomach — and he was newly all stomach — were so swift in their action that whatever he swallowed was digested almost in the swallowing. Since coming upon the ledge he had clutched and devoured two small basking sharks, from six to eight feet long, and a sawfish fully ten feet long, who had not been on their guard against the approach of such a peril. Besides these substantial victims, countless small fry, of every kind, had been drawn deftly to the insatiable vortex of his maw. Nevertheless, his appetite was again crying out. He tried the wreck again, first carefully, then boldly, till the writhing tentacles, with their sensitive tips and suckers, had enveloped it from stem to stern and searched it inside and out. A few lurking fish and molluscs were snatched from the dark interior by those insinuating and inexorable feelers; and a toothsome harvest of anchored crustaceans was gathered from the hidden outfaces along beside the keel. But of the bodies of the pirates that had gone down in the sudden foundering there was nothing left bat bones, which the myriad scavengers of the sea had polished to the barren smoothness of ivory.

While the pallid monster was occupied in the investigation of the wreck those two great bulging black mirrors of his eyes were sleeplessly alert to everything that passed above or about them. Once a swordfish, about seven feet long, sailed carelessly though swiftly some ten feet overhead. Up darted a livid tentacle, and fixed upon it with the deadly sucking discs. In vain the splendid and ferocious fish lashed out in the effort to wrench itself free. In vain it strove to plunge downward and pierce the puffy monster with its sword. In a second or two more tentacles were wrapped about it. Then, all force crushed out of it, it was dragged down and crammed into the conquerors horrible mouth.

While its mouth was yet working with the satisfaction of this meal, the monster saw a graceful but massive black shape, nearly half as long as himself, swimming slowly between his eyes and the shining surface. At the sight a shudder of fear passed over him. Every waving tentacle shrank back and lay moveless, as if suddenly paralysed, and he flattened himself down as best he could beside the dark hulk of the wreck. Well he knew that dark shape was a whale — and a whale was the one being he knew of which he had cause to fear. Against those rending jaws his cable-like tentacles and tearing beak were of no avail, his unarmoured body utterly defenceless.

The whale, however — not a sperm, but one of a much smaller, though more savage species, the "killer" — did not catch sight of the giant cuttlefish cringing below him. Intent on other game, he passed swiftly on. His presence, however, had for the moment destroyed the monster's appetite. Instead of continuing his search for food, he wanted a hiding-place. He could no longer be at ease for a moment there in the open.

Just behind the wreck the rock wall rose abruptly to the surface of the reef. Its base was hollowed into a series of low caves, where masses of softer rock had been eaten out from beneath a slanting stratum of more enduring material. The more spacious of these caves was immediately behind the wreck. It was exactly what the monster craved. He backed into it with alacrity, completely filling it with his spectral and swollen body. In the doorway the convex inky lenses of his eyes kept watch, moveless and all-seeing. And his ten pale-spotted tentacles, each thicker at the base than a man's thigh, lay outspread and hidden among the seaweeds, waiting for such victims as might come within reach of their lightning snap and coil.

The monster had no more than got himself fairly installed in his new quarters, when into the range of his awful eyes came a singular figure, descending slowly through the glimmering green directly over the wreck. It was not so long as the swordfish he had lately swallowed, but it was thick and massive-looking; and it was blunt at the ends, unlike any fish he had ever seen. Its eyes were enormous, round and bulging. From its head and from one of its curious round, thick fins, extended two slender antennæ straight up towards the surface, and so long that their extremities were beyond the monster's vision. It was indeed a strange-looking creature, but he felt sure that it would be very good to eat. In their concealment among the many-coloured seaweeds his tentacles thrilled with expectancy, and he waited, like some stupendous nightmare of a spider, to spring the moment the prey came within reach.

It chanced, however, that just as the strange creature, descending without any movement of its fins, did come within reach, there also appeared again, in the distance, the black form of the "killer" whale, swimming far overhead. The monster changed his plans instantly. His interest in the newcomer died out. He became intent on nothing but keeping himself inconspicuous. The newcomer, unconscious of the terror lying in wait so near him and of the dark form patrolling the upper green, alighted upon the wreck and groped his way lumberingly into the cabin, dragging those two slim antennæ behind him.

IV.

When Jan Laurvik, in his up-to-date and well-tested diving-suit, went down through the green twilight of the sea, he was doing what it was his profession to do, and he had few misgivings. He had confidence in his equipment, in his skill, and in his mate at the rope and the air-pump, Captain Jerry. For defence against any obtrusive shark or sawfish he carried a heavy, long-bladed two-edged knife, by far the most effective weapon in deep water. This knife he wore in a sheath at his waist, with a cord attached to the handle so that it could not get away from him. He carried also a tiny electric battery supplying a strong lamp on the front of his head-piece just above his eyes.

From his long experience in sounding and in locating wrecks, Jan Laurvik had acquired an accuracy that seemed almost like divination. His soundings, in this instance, had been particularly thorough, because he did not wish to waste any more time than necessary at the depth in which he would have to work. He was not surprised, therefore, when he found himself descending upon the wreck of a junk. Moreover, as it was not an old wreck, he concluded that it was the junk which he was looking for. The wreck had settled almost on an even keel; and as he was familiar with craft of her type, he had no difficulty in finding his way about.

It was in the narrow, closet-like structure which served as the junk's cabin that the pirate had said the pearls would be found. The door was open. Turning on his light, which struggled with the water and diffused a ghostly glow, he found himself confronted by a hideous little joss of red-and-gilt lacquer. He knew it was lacquer, and of the best, for nothing else, except gold itself, would have withstood the months of soaking in sea-water. Jan grinned to himself, there within his rubber and copper shell, at this evidence of pirate piety. Then it occurred to him that a man like the pirate captain would probably have turned his piety to practical use. What better guardian of the treasure than a god? Dragging the gaudy deity from his altar, he found the altar hollow. In that secure receptacle lay a series of packages done up with careful precision in wrappings of oiled silk. He knew the style of wrapping very well. For all his coolness his heart fell to thumping painfully at the sight of this vast wealth beneath his hand. Then he realized that the pressure of the water, and of the compressed air in his helmet, was beginning to tell upon him. In fierce but orderly haste he corded the packages about his middle and turned to leave the cabin. He would make another trip for the lacquer god, and for such other articles of value or vertu as the junk might contain.

Jan turned to leave the cabin. But in the doorway he started back with a shudder of dread and loathing. A slender, twisting thing, whitish in colour and minutely speckled with livid spots, reached in, and fastened upon his arm with soft-looking suckers which held like death.

Jan knew instantly what the pale, writhing thing was. Out flashed his knife. With a swift stroke he slashed off the detaining tip, where it had a thickness of perhaps two inches. The raw stump shrank back like a severed worm, and Jan, leaping clear of the doorway, signalled furiously to be hauled up. But at the same instant two more of the curling white things came reaching over the bulwarks and fastened upon him — one upon his right arm, hampering him so that he was almost helpless, and the other upon his left leg just above the knee. He felt his signal promptly answered by a powerful tug on the rope. But he was anchored to the wreck as if he had grown to it.

Never before had Jan Laurvik felt the clutch of fear at his heart as he did at this moment. But not for an instant, in the horror, did he lose his presence of mind. He knew that in a pulling match with the giant devil-fish of the deeps his comrades in the boat far overhead would be nowhere. He had made a mistake in leaving the cabin. Frantically he signalled with his left hand, to "slack away" on the rope; and at the same time, though hampered by the grip on his right arm, he managed to slash off the end of the feeler that had fixed upon his leg. On the instant, whipping the knife over to his left, he cut his right arm clear, and sprang back into the doorway.

Jan's idea was that by keeping just inside the cabin door he could defend himself from being surrounded by the assault of the writhing things. He knew that in the open he would speedily be enfolded and crushed, and engulfed between the jaws of the monstrous squid. But in the narrow doorway the swift play of his blade would have some chance. He gained the doorway. He got fairly inside it, indeed. But as he entered he was horrified to see the thick stump, whose tip he had shorn off, dart in with him and fix itself, by its bigger and more irresistible suckers, upon the middle of his breast. With a shiver he sliced off the fatal discs, in one long sweep of his blade; then turned like a flash to sever a pallid tip which had fastened upon his helmet.

Jan was now thankful enough that he had got himself into the narrow doorway. Seemingly undisturbed by the slashings and slicings which some of them had received, the whole ten squirming horrors now darted at the doorway. Jan's knife swooped this way and that; but as fast as he severed one clutch two more would make good. The cut tentacles grew to be the more terrifying, because their suckers were so big, and they themselves were so thick and hard to cut. Presently no fewer than three of the diabolical things laid their loathsome hold upon his right leg, below the knee, and began to haul it out through the door. Jan slashed at them madly, but not altogether effectually, for at this moment another tentacle had laid grip upon his arm below the elbow. He had just time to shift the knife again to his left and catch the jamb of the door, when he felt his helmet almost jerked from his head. This grip he dared not interfere with, lest he should cut, at the same time, the air-tube that fed his lungs, and drown like a rat in a hole. All he could do was to hold on to the door-jamb, and carve away savagely at the tentacles which were within reach. If he could get free of those, he calculated that he could then reach the one which had fastened to his headpiece by throwing himself over on his back and so bringing it within range of his vision and his knife. At this moment, however, just as the pressure upon his neck was becoming intolerable, he felt his head suddenly released. One of the great sucking discs had crushed in the glass of the electric lamp and fastened upon the live wire. The sensation it experienced was evidently not pleasant, for it let go promptly, and secured a new hold upon Jan's left arm.

This hold left him almost helpless, because he could no longer wield the knife freely with either hand. He felt himself slowly being pulled out of the doorway by his right leg. Throwing himself partly backward, and partly behind the door, he gained a firmer brace and at the same time brought his knife again into better play. He would fight to the very last gasp, but he felt that the odds had now gone overwhelmingly against him. The fear of death itself was not heavy upon him. He had faced it too often, and too coolly, for that. But at the manner of this death that confronted him his very soul sickened with loathing. As he thought of it, his horror was not lessened by the sight which now greeted his view. A colossal, swollen, leprous-looking bulk, pallid and spotted, was mounting over the bulwark. Two great oval lenses of clear blackness, set close together, were in the front of the bulk, just over the spot where the tentacles started. These gigantic, appalling, expressionless eyes were fixed upon him. The monster was coming aboard to see what kind of creature it was that was giving him so much trouble.

Jan saw that the end of the fight was very near. The thought, however, did not unnerve him. Rather, it put new fire into his nerves and muscles. By a tremendous wrench he

ABOVE: Illustration from page 304 of *The Haunters of the Silences: A Book of Animal Life*, illustrated by Charles Livingston Bull, 1907.
© Courtesy of Internet Archive

succeeded in reaching with the knife the tentacle that bound his right arm. This freedom was like a new lease of life to him. He made swift play with his blade, so savagely that

ABOVE: Photograph of Charles GD Roberts, *c.* 1890.
© Courtesy of Esther Clark Archives, Acadia University, Harry Starr accession no. 1990-00

he was able to drag himself back almost completely into the cabin before the writhing horrors again closed upon him. But meanwhile the monster's gigantic body had gained the deck. Those two awful eyes were slowly drawing nearer; and below them he saw the viscid mouth opening and shutting in anticipation.

At this a kind of madness began to surge up in Jan Laurvik's overtaxed brain. His veins seemed to surge with fresh power, as if there were nothing too tremendous for him to accomplish. He was on the very point of stopping his resistance, plunging straight in among the arms, and burying his big blade in those unspeakable eyes. It would be a satisfaction, at least, to force them to change their expression. And then, well something might happen!

Before he could put this desperate scheme into execution, however, something did happen. Jan was aware of a sudden darkness overhead. The monster was evidently aware of it too, for every one of the twisting horrors suddenly shrank away, leaving Jan to lean up against the doorway, free. The next moment, a huge black shape descended perpendicularly upon the fleshy mountain of the monster's back, and a rush of water drove Jan backward into the cabin.

As the electric lamp had gone out when the glass was broken, Jan could see but dimly the awful battle of giants now going on before him. So excited was he that he forgot his own new peril. The danger was now that in the struggle one or other of the battling bulks might well crush the cabin' flat, or entangle the air-tube and life-line. In

160 'The Terror of the Sea Caves' by Sir Charles GD Roberts

either case Jan's finish would be swift; but in comparison with the loathsome death from which he had just been so miraculously saved, such an end seemed no very dreadful thing. He was altogether absorbed in watching the prowess of his avenging rescuer.

Skilled in deep-sea lore as he was, he knew the dark fury which had swooped down upon the devil-fish. It was a "killer" whale, or grampus, the most redoubtable and implacable fighter of all the kindreds of the sea. Jan saw its wide jaws shear off three mighty tentacles at once, close at the base. The others writhed up hideously and fastened upon him, but under the surging of his resistless muscles their tissues tore apart like snapped cables. Huge masses of the monster's ghastly flesh were bitten off, and thrown aside. Then, gaining a grip that took in the monster's head and the roots of the tentacles, the killer shook his prey as a bulldog might shake a fat sheep. The tentacles straightened out sharply. Jan saw that the fight was over; and that it was high time for him to remove from that too strenuous neighbourhood. He gave the signal vehemently, and was drawn up without attracting his dangerous rescuer's notice. When Captain Jerry hauled him in over the boatside, he fell in an unconscious heap.

When Jan came to himself he was in his bunk on the *Sarawak*. It was an utter physical and nervous exhaustion that had overcome him. His swoon had passed into a heavy sleep, and when he awoke he sat up with a start. Captain Jerry was at his side, bursting with suppressed curiosity; and the Scottish engineer was standing by the bunk.

"Waal, partner, you've delivered the goods all right!" drawled Captain Jerry. "They're the stuff; not a doubt of it. But kind o' seemed to us up here you were having high jinks of one kind or another down there. What was it?"

"It was hell!" responded Jan with a shudder. Then he took hold of Captain Jerry's hand, and felt it as if to make sure it was real, or as if he needed the feel of honest human flesh again to bring him to his senses.

"Ugh!" he went on, swinging out of the bunk. "Let me get out into the sunlight again! Let me see the sky again! I'll tell you all about it by-an'-by, Jerry. But wait. Were all the packages on me, all right?"

"I reckon!" responded Captain Jerry. "There was six of 'em tied on to you. I reckon they're worth the three hundred an' fifty thousand all right!"

"Well, let's get away from this place quick as we can get steam up again!" said Jan. "There's more swag down there, I guess — lots of it. But I wouldn't go down again, nor send another man down, for all the millions we've all of us ever heard tell of. Mr McWha, how soon can we be moving?"

"Ten meenutes, more or less!" replied the Scotsman.

"All right! When we're outside of this accursed bay, an' round the 'Camel' yonder, I'll tell you what it's like down there under that shiny green."

'THE SEAFARER'
BY EZRA POUND
(1911)

Ezra Weston Loomis Pound (1885–1972) was born in Hailey, Idaho, but lived most of his adult life in Europe. In London, he immersed himself in the literary scene and, along with Hilda Doolittle (H.D.) and Richard Aldington, developed the Imagist movement in poetry (see H.D., page 179). After the First World War, which Pound believed was caused by a Jewish financial conspiracy, his antisemitism intensified. He moved to Paris and began writing work in praise of Benito Mussolini's fascist Italy. During the Second World War, he recorded broadcasts for the Republic of Salò and was a supporter of Nazi eugenics.

In 1945, the US government arrested Pound for treason, but he was declared mentally unfit for trial. He remained an inpatient at St Elizabeth's Hospital in Washington, DC until 1958 but continued to write fascistic essays and poems, which his friends attributed to insanity. Pound then lived in Italy until his death at the age of 87.

'The Seafarer' is a loose recreation of a poem in the Exeter Book, a 10th-century manuscript of Old English poetry. Pound's version was first published in *The New Age* in 1911, in a column titled 'I Gather the Limbs of Osiris'. While Pound reproduces the alliterative verse and strong stresses of the original, it is not otherwise considered a linguistically accurate translation, containing only 99 of the original 124 lines. The value in Pound's 'The Seafarer' lies instead in his application of the Modernist edict 'Make it New', which itself is the title of Pound's 1934 collection of essays on breaking from established form and tradition.

LEFT: The galleon *La Nostra Sengnora* setting out from Sanlucar for Venezuela, from Jerome Coeler's account of his travels in 1533–34.
© Courtesy of the British Library/Bridgeman Images

'The Seafarer'

May I for my own self song's truth reckon,
Journey's jargon, how I in harsh days
Hardship endured oft.
Bitter breast-cares have I abided,
Known on my keel many a care's hold,
And dire sea-surge, and there I oft spent
Narrow nightwatch nigh the ship's head
While she tossed close to cliffs. Coldly afflicted,
My feet were by frost benumbed.
Chill its chains are; chafing sighs
Hew my heart round and hunger begot
Mere-weary mood. Lest man know not
That he on dry land loveliest liveth,
List how I, care-wretched, on ice-cold sea,
Weathered the winter, wretched outcast
Deprived of my kinsmen;
Hung with hard ice-flakes, where hail-scur flew,
There I heard naught save the harsh sea
And ice-cold wave, at whiles the swan cries,
Did for my games the gannet's clamour,
Sea-fowls, loudness was for me laughter,
The mews' singing all my mead-drink.
Storms, on the stone-cliffs beaten, fell on the stern
In icy feathers; full oft the eagle screamed
With spray on his pinion.
Not any protector
May make merry man faring needy.
This he little believes, who aye in winsome life
Abides 'mid burghers some heavy business,
Wealthy and wine-flushed, how I weary oft
Must bide above brine.
Neareth nightshade, snoweth from north,
Frost froze the land, hail fell on earth then
Corn of the coldest. Nathless there knocketh now
The heart's thought that I on high streams
The salt-wavy tumult traverse alone.
Moaneth alway my mind's lust
That I fare forth, that I afar hence
Seek out a foreign fastness.
For this there's no mood-lofty man over
    earth's midst,
Not though he be given his good, but will have
    in his youth greed;
Nor his deed to the daring, nor his king to
    the faithful
But shall have his sorrow for sea-fare
Whatever his lord will.
He hath not heart for harping, nor in ring-having
Nor winsomeness to wife, nor world's delight
Nor any whit else save the wave's slash,
Yet longing comes upon him to fare forth on
    the water.
Bosque taketh blossom, cometh beauty of berries,
Fields to fairness, land fares brisker,
All this admonisheth man eager of mood,
The heart turns to travel so that he then thinks
On flood-ways to be far departing.
Cuckoo calleth with gloomy crying,
He singeth summerward, bodeth sorrow,
The bitter heart's blood. Burgher knows not —
He the prosperous man — what some perform
Where wandering them widest draweth.
So that but now my heart burst from my breast-lock,
My mood 'mid the mere-flood,

ABOVE: Portrait of Ezra Pound, illustrated by Wyndham Lewis, 1919.
© Courtesy of Wikimedia Commons

Over the whale's acre, would wander wide.
On earth's shelter cometh oft to me,
Eager and ready, the crying lone-flyer,
Whets for the whale-path the heart irresistibly,
O'er tracks of ocean; seeing that anyhow
My lord deems to me this dead life
On loan and on land, I believe not
That any earth-weal eternal standeth
Save there be somewhat calamitous
That, ere a man's tide go, turn it to twain.

Disease or oldness or sword-hate
Beats out the breath from doom-gripped body.
And for this, every earl whatever, for those
    speaking after —
Laud of the living, boasteth some last word,
That he will work ere he pass onward,
Frame on the fair earth 'gainst foes his malice,
Daring ado, ...
So that all men shall honour him after
And his laud beyond them remain 'mid the English,
Aye, for ever, a lasting life's-blast,
Delight mid the doughty.
Days little durable,
And all arrogance of earthen riches,
There come now no kings nor Cæsars
Nor gold-giving lords like those gone.
Howe'er in mirth most magnified,
Whoe'er lived in life most lordliest,
Drear all this excellence, delights undurable!
Waneth the watch, but the world holdeth.
Tomb hideth trouble. The blade is layed low.
Earthly glory ageth and seareth.
No man at all going the earth's gait,
But age fares against him, his face paleth,
Grey-haired he groaneth, knows gone companions,
Lordly men are to earth o'ergiven,
Nor may he then the flesh-cover, whose life ceaseth,
Nor eat the sweet nor feel the sorry,
Nor stir hand nor think in mid heart,
And though he strew the grave with gold,
His born brothers, their buried bodies
Be an unlikely treasure hoard.

'The Seafarer' by Ezra Pound

'THE THING IN THE WEEDS'

BY WILLIAM HOPE HODGSON
(1913)

William Hope Hodgson (1877–1918) was a writer, sailor, soldier, bodybuilder and photographer. Born in Essex, England, he ran away from boarding school at 13 to become a cabin boy and spent his teenage years as a sailor. This was a largely negative experience, a sentiment reflected across his writing. In 1899 he opened W. H. Hodgson's School of Physical Culture in Blackburn, and published papers on training and exercise. He was killed, aged 40, in the Fourth Battle of Ypres in the First World War.

Hodgson published his first short story, 'The Goddess of Death', in 1904 and his first novel, *The Boats of the "Glen Carrig"*, in 1907, and continued to write extensively throughout his life. His works contain elements of cosmic horror, science fiction and the fantastic. In his 1927 essay 'Supernatural Horror in Literature', HP Lovecraft identifies in Hodgson's work 'vast occasional power in its suggestion of lurking worlds and beings behind the ordinary surface of life', and names him a prominent writer of 'the weird short story'.

Hodgson returned to the sea often in his stories. Many of his works are set in the Sargasso Sea. The title of 'The Thing in the Weeds' (1913) references the *Sargassum* algae for which this region is named, and speaks to the fear of unseen depths.

LEFT: A sea monster attacking a ship, from *Terra Sancta*, Abraham Ortelius, 1585.
© Courtesy of Wikimedia Commons

'The Thing in the Weeds'

I.

This is an extraordinary tale. We had come up from the Cape, and owing to the Trades heading us more than usual, we had made some hundreds of miles more westing than I ever did before or since.

I remember the particular night of the happening perfectly. I suppose what occurred stamped it solid into my memory, with a thousand little details that, in the ordinary way, I should never have remembered an hour. And, of course, we talked it over so often among ourselves that this, no doubt, helped to fix it all past any forgetting.

I remember the mate and I had been pacing the weather side of the poop and discussing various old shellbacks' superstitions. I was third mate, and it was between four and five bells in the first watch, i.e. between ten and half-past.

Suddenly he stopped in his walk and lifted his head and sniffed several times.

"My word, mister," he said, "there's a rum kind of stink somewhere about. Don't you smell it?"

I sniffed once or twice at the light airs that were coming in on the beam; then I walked to the rail and leaned over, smelling again at the slight breeze. And abruptly I got a whiff of it, faint and sickly, yet vaguely suggestive of something I had once smelt before.

"I can smell something, Mr. Lammart," I said. "I could almost give it name; and yet somehow I can't." I stared away into the dark to windward. "What do you seem to smell?" I asked him.

"I can't smell anything now," he replied, coming over and standing beside me. "It's gone again. No! By Jove! there it is again. My goodness! Phew!"

The smell was all about us now, filling the night air. It had still that indefinable familiarity about it, and yet it was curiously strange, and, more than anything else, it was certainly simply beastly.

The stench grew stronger, and presently the mate asked me to go for'ard and see whether the look-out man noticed anything. When I reached the break of the fo'c's'le head I called up to the man, to know whether he smelled anything.

"Smell anythin', sir?" he sang out. "Jumpin' larks! I sh'u'd think I do. I'm fair p'isoned with it."

I ran up the weather steps and stood beside him. The smell was certainly very plain up there, and after savouring it for a few moments I asked him whether he thought it might be a dead whale. But he was very emphatic that this could not be the case, for, as he said, he had been nearly fifteen years in whaling ships, and knew the

RIGHT: *Aquatilium Animalium Historiæ liber primus* (squid), illustrated by Ippolito Salviani, 1554. © Courtesy of Wikimedia Commons

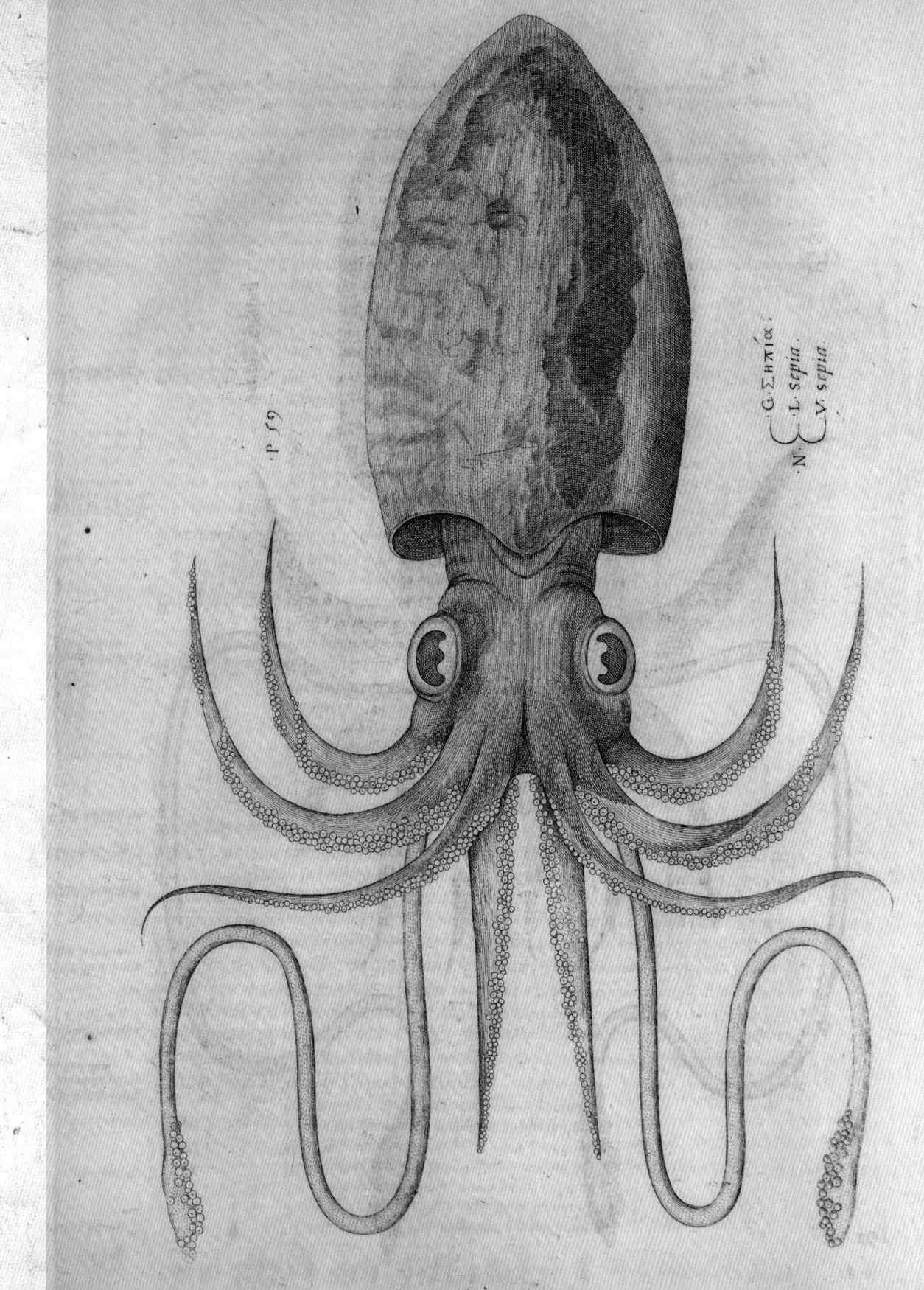

G. Σηπία.
L. sepia
V. sepia

smell of a dead whale, "like as you would the smell of bad whiskey, sir," as he put it. " 'Tain't no whale yon, but the Lord He knows what 'tis. I'm thinking it's Davy Jones come up for a breather."

I stayed with him some minutes, staring out into the darkness, but could see nothing; for, even had there been something big close to us, I doubt whether I could have seen it, so black a night it was, without a visible star, and with a vague, dull haze breeding an indistinctness all about the ship.

I returned to the mate and reported that the look-out complained of the smell, but that neither he nor I had been able to see anything in the darkness to account for it.

By this time the queer, disgusting odour seemed to be in all the air about us, and the mate told me to go below and shut all the ports, so as to keep the beastly smell out of the cabins and the saloon.

When I returned he suggested that we should shut the companion doors, and after that we commenced to pace the poop again, discussing the extraordinary smell, and stopping from time to time to stare through our night-glasses out into the night about the ship.

"I'll tell you what it smells like, mister," the mate remarked once, "and that's like a mighty old derelict I once went aboard in the North Atlantic. She was a proper old-timer, an' she gave us all the creeps. There was just this funny, dank, rummy sort of century-old bilge-water and dead men an' seaweed. I can't stop thinkin' we're nigh some lonesome old packet out there; an' a good thing we've not much way on us!"

"Do you notice how almighty quiet everything's gone the last half-hour or so?" I said a little later. "It must be the mist thickening down."

"It is the mist," said the mate, going to the rail and staring out. "Good Lord, what's that?" he added.

Something had knocked his hat from his head, and it fell with a sharp rap at my feet. And suddenly, you know, I got a premonition of something horrid.

"Come away from the rail, sir!" I said sharply, and gave one jump and caught him by the shoulders and dragged him back. "Come away from the side!"

"What's up, mister?" he growled at me, an twisted his shoulders free. "What's wrong with you? Was it you knocked off my cap?" He stooped and felt around for it, and as he did so I heard something unmistakably fiddling away at the rail which the mate had just left.

"My God, sir!" I said "there's something there. Hark!"

The mate stiffened up, listening; then he heard it. It was for all the world as if something was feeling and rubbing the rail there in the darkness, not two fathoms away from us.

"Who's there?" said the mate quickly. Then, as there was no answer: "What the devil's this hanky-panky? Who's playing the goat there?" He made a swift step through the darkness towards he rail, but I caught him by the elbow.

"Don't go, mister!" I said, hardly above a whisper. "It's not one of the men. Let me get a light."

"Quick, then!" he said, and I turned and ran aft to the binnacle and snatched out the lighted lamp. As I did so I heard the mate shout something out of the darkness in a strange voice. There came a sharp, loud,

170   'The Thing in the Weeds' by William Hope Hodgson

rattling sound, and then a crash, and immediately the mate roaring to me to hasten with the light. His voice changed even whilst he shouted, and gave out something that was nearer a scream than anything else. There came two loud, dull blows and an extraordinary gasping sound; and then, as I raced along the poop, there was a tremendous smashing of glass and an immediate silence.

"Mr. Lammart!" I shouted. "Mr. Lammart!" And then I had reached the place where I had left the mate for forty seconds before; but he was not there.

"Mr. Lammart!" I shouted again, holding the light high over my head and turning quickly to look behind me. As I did so my foot glided on some slippery substance, and I went headlong to the deck with a tremendous thud, smashing the lamp and putting out the light.

I was on my feet again in an instant. I groped a moment for the lamp, and as I did so I heard the men singing out from the maindeck and the noise of their feet as they came running aft. I found the broken lamp and realised it was useless; then I jumped for the companion-way, and in half a minute I was back with the big saloon lamp glaring bright in my hands.

I ran for'ard again, shielding the upper edge of the glass chimney from the draught of my running, and the blaze of the big lamp seemed to make the weather side of the poop as bright as day, except for the mist, that gave something of a vagueness to things.

Where I had left the mate there was blood upon the deck, but nowhere any signs of the man himself. I ran to the weather rail and held the lamp to it. There was blood upon it, and the rail itself seemed to have been wrenched by some huge force. I put out my hand and found that I could shake it. Then I leaned out-board and held the lamp at arm's length, staring down over the ship's side.

"Mr. Lammart!" I shouted into the night and the thick mist. "Mr. Lammart! Mr. Lammart!" But my voice seemed to go, lost and muffled and infinitely small, away into the billowy darkness.

I heard the men snuffling and breathing, waiting to leeward of the poop. I whirled round to them, holding the lamp high,

"We heard somethin', sir," said Tarpley, the leading seaman in our watch. "Is anythin' wrong, sir?"

"The mate's gone," I said blankly. "We heard something, and I went for the binnacle lamp. Then he shouted, and I heard a sound of things smashing, and when I got back he'd gone clean." I turned and held the light out again over the unseen sea, and the men crowded round along the rail and stared, bewildered.

"Blood, sir," said Tarpley, pointing. "There's somethin' almighty queer out there." He waved a huge hand into the darkness. "That's what stinks —"

He never finished; for suddenly one of the men cried out something in a frightened voice: "Look out, sir! Look out, sir!"

I saw, in one brief flash of sight, something come in with an infernal flicker of movement; and then, before I could form any notion of what I had seen, the lamp was dashed to pieces across the poop deck. In that instant my perceptions cleared, and I saw the incredible folly of what we were doing; for there we were, standing up against the blank, unknowable night, and out there in the darkness

'The Thing in the Weeds' by William Hope Hodgson

there surely lurked some thing of monstrousness; and we were at its mercy. I seemed to feel it hovering—hovering over us, so that I felt the sickening creep of gooseflesh all over me.

"Stand back from the rail!" I shouted. "Stand back from the rail!" There was a rush of feet as the men obeyed, in sudden apprehension of their danger, and I gave back with them. Even as I did so I felt some invisible thing brush my shoulder, and an indescribable smell was in my nostrils from something that moved over me in the dark.

"Down into the saloon everyone!" I shouted. "Down with you all! Don't wait a moment!"

There was a rush along the dark weather deck, and then the men went helter-skelter down the companion steps into the saloon, falling and cursing over one another in the darkness. I sang out to the man at the wheel to join them, and then I followed.

I came upon the men huddled at the foot of the stairs and filling up the passage, all crowding each other in the darkness. The skipper's voice was filling the saloon, and he was demanding in violent adjectives the cause of so tremendous a noise. From the steward's berth there came also a voice and the splutter of a match, and then the glow of a lamp in the saloon itself.

I pushed my way through the men and found the captain in the saloon in his sleeping gear, looking both drowsy and angry, though perhaps bewilderment topped every other feeling. He held his cabin lamp in his hand, and shone the light over the huddle of men.

I hurried to explain, and told him of the incredible disappearance of the mate, and of my conviction that some extraordinary thing was lurking near the ship out in the mist and the darkness. I mentioned the curious smell, and told how the mate had suggested that we had drifted down near some old-time, sea-rotted derelict. And, you know, even as I put it into awkward words, my imagination began to awaken to horrible discomforts; a thousand dreadful impossibilities of the sea became suddenly possible.

The captain (Jeldy was his name) did not stop to dress, but ran back into his cabin, and came out in a few moments with a couple of revolvers and a handful of cartridges. The second mate had come running out of his cabin at the noise, and had listed intensely to what I had to say; and now he jumped back into his berth and brought out his own lamp and a large Smith and Wesson, which was evidently ready loaded.

Captain Jeldy pushed one of his revolvers into my hands, with some of the cartridges, and we began hastily to load the weapons. Then the captain caught up his lamp and made for the stairway, ordering the men into the saloon out of his way.

"Shall you want them, sir?" I asked.

"No," he said. "It's no use their running any unnecessary risks." He threw a word over his shoulder: "Stay quiet here, men; if I want you I'll give you a shout; then come spry!"

"Aye, aye, sir," said the watch

LEFT: 'At last he came to the great sea-serpent himself, lying dead at the bottom.' from page 232 of *The Water-Babies*, illustrated by Warwick Goble, 1909.
© Courtesy of Alamy

'The Thing in the Weeds' by William Hope Hodgson

in a chorus; and then I was following the captain up the stairs, with the second mate close behind.

We came up through the companion-way on to the silence of the deserted poop. The mist had thickened up, even during the brief time that I had been below, and there was not a breath of wind. The mist was so dense that it seemed to press in upon us, and the two lamps made a kind of luminous halo in the mist, which seemed to absorb their light in a most peculiar way.

"Where was he?" the captain asked me, almost in a whisper.

"On the port side, sir," I said, "a little foreside the charthouse and about a dozen feet in from the rail. I'll show you the exact place."

We went for'ard along what had been the weather side, going quietly and watchfully, though, indeed, it was little enough that we could see, because of the mist. Once, as I led the way, I thought I heard a vague sound somewhere in the mist, but was all unsure because of the slow creak, creak of the spars and gear as the vessel rolled slightly upon an odd, oily swell. Apart from this slight sound, and the far-up rustle of the canvas slatting gently against the masts, there was no sound of all throughout the ship. I assure you the silence seemed to me to be almost menacing, in the tense, nervous state in which I was.

"Hereabouts is where I left him," I whispered to the captain a few seconds later. "Hold your lamp low, sir. There's blood on the deck."

Captain Jeldy did so, and made a slight sound with his mouth at what he saw. Then, heedless of my hurried warning, he walked across to the rail, holding his lamp high up. I followed him, for I could not let him go alone; and the second mate came too, with his lamp. They leaned over the port rail and held their lamps out into the mist and the unknown darkness beyond the ship's side. I remember how the lamps made just two yellow glares in the mist, ineffectual, yet serving somehow to make extraordinarily plain the vastitude of the night and the possibilities of the dark. Perhaps that is a queer way to put it, but it gives you the effect of that moment upon my feelings. And all the time, you know, there was upon me the brutal, frightening expectancy of something reaching in at us from out of that everlasting darkness and mist that held all the sea and the night, so that we were just three mist-shrouded, hidden figures, peering nervously.

The mist was now so thick that we could not even see the surface of the water overside, and fore and aft of us the rail vanished away into the fog and the dark. And then, as we stood here staring, I heard something moving down on the maindeck. I caught Captain Jeldy by the elbow.

"Come away from the rail, sir," I said, hardly above a whisper; and he, with the swift premonition of danger, stepped back and allowed me to urge him well inboard. The second mate followed, and the three of us stood there in the mist, staring round about us and holding our revolvers handily, and the dull waves of the mist beating in slowly upon the lamps in vague wreathings and swirls of fog.

"What was it you heard, mister?" asked the captain after a few moments.

"Ssst!" I muttered. "There it is again. There's something

moving down on the maindeck!"

Captain Jeldy heard it himself now, and the three of us stood listening intensely. Yet it was hard to know what to make of the sounds. And then suddenly there was the rattle of a deck ringbolt, and then again, as if something or someone were fumbling and playing with it.

"Down there on the maindeck!" shouted the captain abruptly, his voice seeming hoarse close to my ear, yet immediately smothered by the fog. "Down there on the maindeck! Who's there?"

But there came never an answering sound. And the three of us stood there, looking quickly this way and that, and listening. Abruptly the second mate muttered something:

"The look-out, sir! The look-out!"

Captain Jeldy took the hint on the instant.

"On the look-out there!" he shouted.

And then, far away and muffled-sounding, there came the answering cry of the look-out man from the fo'c'sle head:

"Sir-r-r?" A little voice, long drawn out through unknowable alleys of fog.

"Go below into the fo'c'sle and shut both doors, an' don't stir out till you're told!" sung out Captain Jeldy, his voice going lost into the mist. And then the man's answering "Aye, aye, sir!" coming to us faint and mournful. And directly afterwards the clang of a steel door, hollow-sounding and remote; and immediately the sound of another.

"That puts them safe for the present, anyway," said the second mate. And even as he spoke there came again that indefinite noise down upon the maindeck of something moving with an incredible and unnatural stealthiness.

"On the maindeck there!" shouted Captain Jeldy sternly. "If there is anyone there, answer, or I shall fire!"

The reply was both amazing and terrifying, for suddenly a tremendous blow was stricken upon the deck, and then there came the dull, rolling sound of some enormous weight going hollowly across the maindeck. And then an abominable silence.

"My God!' said Captain Jeldy in a low voice, "what was that?" And he raised his pistol, but I caught him by the wrist. "Don't shoot, sir!" I whispered. "It'll be no good. That—that—whatever it is I—mean it's something enormous, sir. I—I really wouldn't shoot." I found it impossible to put my vague idea into words; but I felt there was a force aboard, down on the maindeck, that it would be futile to attack with so ineffectual a thing as a puny revolver bullet.

And then, as I held Captain Jeldy's wrist, and he hesitated, irresolute, there came a sudden bleating of sheep and the sound of lashings being burst and the cracking of wood; and the next instant a huge crash, followed by crash after crash, and the anguished m-aa-a-a-ing of sheep.

"My God!" said the second mate, "the sheep-pen's being beaten to pieces against the deck. Good God! What sort of thing could do that?"

The tremendous beating ceased, and there was a splashing overside; and after that a silence so profound that it seemed as if the whole atmosphere of the night was full of an unbearable, tense quietness. And then the damp slatting of a sail, far up in the night, that made me start—a lonesome sound to break suddenly through that infernal silence upon my raw nerves.

"Get below, both of you. Smartly now!" muttered Captain Jeldy. "There's something run either aboard us or alongside; and we can't do anything till daylight."

We went below and shut the doors of the companion-way, and there we lay in the wide Atlantic, without wheel or look-out or officer in charge, and something incredible down on the dark maindeck.

II.

For some hours we sat in the captain's cabin talking the matter over whilst the watch slept, sprawled in a dozen attitudes on the floor of the saloon. Captain Jeldy and the second mate still wore their pyjamas, and our loaded revolvers lay handy on the cabin table. And so we watched anxiously through the hours for the dawn to come in.

As the light strengthened we endeavoured to get some view of the sea from the ports, but the mist was so thick about us that it was exactly like looking out into a grey nothingness, that became presently white as the day came.

"Now," said Captain Jeldy, "we're going to look into this."

He went out through the saloon to the companion stairs. At the top he opened the two doors, and the mist rolled in on us, white and impenetrable. For a little while we stood there, the three of us, absolutely silent and listening, with our revolvers handy; but never a sound came to us except the odd, vague slatting of a sail or the slight creaking of the gear as the ship lifted on some slow, invisible swell.

Presently the captain stepped cautiously out on to the deck; he was in his cabin slippers, and therefore made no sound. I was wearing gum-boots, and followed him silently, and the second mate after me in his bare feet. Captain Jeldy went a few paces along the deck, and the mist hid him utterly. "Phew!" I heard him mutter, "the stink's worse than ever!" His voice came odd and vague to me through the wreathing of the mist.

"The sun'll soon eat up all this fog," said the second mate at my elbow, in a voice little above a whisper.

We stepped after the captain, and found him a couple of fathoms away, standing shrouded in the mist in an attitude of tense listening.

"Can't hear a thing!" he whispered. "We'll go for'ard to the break, as quiet as you like. Don't make a sound."

We went forward, like three shadows, and suddenly Captain Jeldy kicked his shin against something and pitched headlong over it, making a tremendous noise. He got up quickly, swearing grimly, and the three of us stood there in silence, waiting lest any infernal thing should come upon us out of all that white invisibility. Once I felt sure I saw something coming towards me, and I raised my revolver, but saw in a moment that there was nothing. The tension of imminent, nervous expectancy eased from us, and Captain Jeldy stooped over the object on the deck.

"The port hencoop's been shifted out here!" he muttered. "It's all stove!"

"That must be what I heard last night when the mate went," I whispered. "There was a loud crash just before he sang out to me to hurry with the lamp."

Captain Jeldy left the smashed hencoop, and the three of us tiptoed silently to the rail across the break of the poop.

Here we leaned over and stared down into the blank whiteness of the mist that hid everything.

"Can't see a thing," whispered the second mate; yet as he spoke I could fancy that I heard a slight, indefinite, slurring noise somewhere below us; and I caught them each by an arm to draw them back.

"There's something down there," I muttered. "For goodness' sake come back from the rail."

We gave back a step or two, and then stopped to listen; and even as we did so there came a slight air playing through the mist.

"The breeze is coming," said the second mate. "Look, the mist is clearing already."

He was right. Already the look of white impenetrability had gone, and suddenly we could see the corner of the after-hatch coamings through the thinning fog. Within a minute we could see as far for'ard as the mainmast, and then the stuff blew away from us, clear of the vessel, like a great wall of whiteness, that dissipated as it went.

"Look!" we all exclaimed together. The whole of the vessel was now clear to our sight; but it was not at the ship herself that we looked, for, after one quick glance along the empty maindeck, we had seen something beyond the ship's side. All around the vessel there lay a submerged spread of weed, for, maybe, a good quarter of a mile upon every side.

"Weed!" sang out Captain Jeldy in a voice of comprehension. "Weed! Look! By Jove, I guess I know now what got the mate!"

He turned and ran to the port side and looked over. And suddenly he stiffened and beckoned silently over his shoulder to us to come and see.

LEFT: Photograph of William Hope Hodgson.
© Courtesy of Wikimedia Commons

'The Thing in the Weeds' by William Hope Hodgson 177

We had followed, and now we stood, one on each side of him, staring.

"Look!" whispered the captain, pointing. "See the great brute! Do you see it? There! Look!"

At first I could see nothing except the submerged spread of the weed, into which we had evidently run after dark. Then, as I stared intently, my gaze began to separate from the surrounding weed a leathery-looking something that was somewhat darker in hue than the weed itself.

"My God!" said Captain Jeldy. "What a monster! What a monster! Just look at the brute! Look at the thing's eyes! That's what got the mate. What a creature out of hell itself!"

I saw it plainly now; three of the massive feelers lay twined in and out among the clumpings of the weed; and then, abruptly, I realised that the two extraordinary round disks, motionless and inscrutable, were the creature's eyes, just below the surface of the water. It appeared to be staring, expressionless, up at the steel side of the vessel. I traced, vaguely, the shapeless monstrosity of what must be termed its head. "My God!" I muttered. "It's an enormous squid of some kind! What an awful brute! What — "

The sharp report of the captain's revolver came at that moment. He had fired at the thing, and instantly there was a most awful commotion alongside. The weed was hove upward, literally in tons. An enormous quantity was thrown aboard us by the thrashing of the monster's great feelers. The sea seemed almost to boil, in one great cauldron of weed and water, all about the brute, and the steel side of the ship resounded with the dull, tremendous blows that the creature gave in its struggle. And into all that whirling boil of tentacles, weed, and seawater the three of us emptied our revolvers as fast as we could fire and reload. I remember the feeling of fierce satisfaction I had in thus aiding to avenge the death of the mate.

Suddenly the captain roared out to us to jump back, and we obeyed on the instant. As we did so the weed rose up into a great mound over twenty feet in height, and more than a ton of it slopped aboard. The next instant three of the monstrous tentacles came in over the side, and the vessel gave a slow, sullen roll to port as the weight came upon her, for the monster had literally hove itself almost free of the sea against our port side, in one vast, leathery shape, all wreathed with weed-fronds, and seeming drenched with blood and curious black liquid.

The feelers that had come inboard thrashed round here and there, and suddenly one of them curled in the most hideous, snake-like fashion around the base of the mainmast. This seemed to attract it, for immediately it curled the two others about the mast, and forthwith wrenched upon it with such hideous violence that the whole towering length of spars, through all their height of a hundred and fifty feet, were shaken visibly, whilst the vessel herself vibrated with the stupendous efforts of the brute.

"It'll have the mast down, sir!" said the second mate, with a gasp. "My God! It'll strain her side open! My — "

"One of those blasting cartridges!" I said to Jeldy almost in a shout, as the inspiration took me. "Blow the brute to pieces!"

"Get one, quick!" said the

captain, jerking his thumb towards the companion. "You know where they are."

In thirty seconds I was back with the cartridge. Captain Jeldy took out his knife and cut the fuse dead short; then, with a steady hand, he lit the fuse, and calmly held it, until I backed away, shouting to him to throw it, for I knew it must explode in another couple of seconds.

Captain Jeldy threw the thing like one throws a quoit, so that it fell into the sea just on the outward side of the vast bulk of the monster. So well had he timed it that it burst, with a stunning report, just as it struck the water. The effect upon the squid was amazing. It seemed literally to collapse. The enormous tentacles released themselves from the mast and curled across the deck helplessly, and were drawn inertly over the rail, as the enormous bulk sank away from the ship's side, out of sight, into the weed. The ship rolled slowly to starboard, and then steadied. "Thank God!" I muttered, and looked at the two others. They were pallid and sweating, and I must have been the same.

"Here's the breeze again," said the second mate, a minute later. "We're moving." He turned, without another word, and raced aft to the wheel, whilst the vessel slid over and through the weedfield.

"Look where that brute broke up the sheep-pen!" cried Jeldy, pointing. "And here's the skylight of the sail-locker smashed to bits!"

He walked across to it, and glanced down. And suddenly he let out a tremendous shout of astonishment:

"Here's the mate down here!" he shouted. "He's not overboard at all! He's here!"

He dropped himself down through the skylight on to the sails, and I after him; and, surely, there was the mate, lying all huddled and insensible on a hummock of spare sails. In his right hand he held a drawn sheath-knife, which he was in the habit of carrying A. B. fashion, whilst his left hand was all caked with dried blood, where he had been badly cut. Afterwards, we concluded he had cut himself in slashing at one of the tentacles of the squid, which had caught him round the left wrist, the tip of the tentacle being still curled tight about his arm, just as it had been when he hacked it through. For the rest, he was not seriously damaged, the creature having obviously flung him violently away through the framework of the skylight, so that he had fallen in a studded condition on to the pile of sails.

We got him on deck, and down into his bunk, where we left the steward to attend to him. When we returned to the poop the vessel had drawn clear of the weed-field, and the captain and I stopped for a few moments to stare astern over the taffrail.

As we stood and looked something wavered up out of the heart of the weed—a long, tapering, sinous thing, that curled and wavered against the dawn-light, and presently sank back again into the demure weed—a veritable spider of the deep, waiting in the great web that Dame Nature had spun for it in the eddy of her tides and currents.

And we sailed away northwards, with strengthening "trades," and left that patch of monstrousness to the loneliness of the sea.

'SEA-HEROES', 'THETIS' AND 'AT ITHACA'

BY H.D.
(1920, 1923 AND 1924)

The American Modernist poet Hilda Doolittle (1886–1961) was born in Bethlehem, Pennsylvania, and studied Greek literature at Bryn Mawr College. In 1911 she travelled to London, where she met her childhood friend Ezra Pound (see page 163). After an on-off relationship with Pound, she married writer and poet Richard Aldington. The three of them lived on the same street in Kensington, London, and together developed the founding principles of the poetic Imagist movement, which was defined by Pound in 'A Retrospect' from his collection of essays on poetry, *Pavannes and Divagations* (1918):

'1. Direct treatment of the "thing" whether subjective or objective.

2. To use absolutely no word that does not contribute to the presentation.

3. As regarding rhythm: to compose in the sequence of the musical phrase, not in sequence of a metronome.'

Pound applied this imagist philosophy to Doolittle's name, which became H.D. in *Des Imagistes, An Anthology* (1914), edited by Pound. H.D.'s imagist poems draw extensively on Greek mythology. The following three poems – 'Sea-Heroes' (from *Coterie*, 1920), 'Thetis' (first published in *Poetry: A Magazine of Verse*, 1923) and 'At Ithaca' (from *Heliodora and Other Poems*, 1924) – reference the sea-deities and monsters of the *Iliad* and *Odyssey* (see page 19).

LEFT: *Design for a Plate with Thetis on a Shell in a Medallion Bordered by Sea Monsters*, engraved by Adriaen Collaert, published by Phillips Galle, *c.* 1600.
© Courtesy of The Met Museum, 17.80.2(2)

'Sea-Heroes'

Crash on crash of the sea,
straining to wreck men;
    sea-boards, continents,
raging against the world, furious,
stay at last, for against your fury
and your mad fight,
the line of heroes stands,
    godlike:
Akroneos, Oknolos, Elatreus,
helm-of-boat, loosener-of-helm,
    dweller-by-sea,
Nauteus, sea-man,
Prumneos, stern-of-ship,
Agchilalos, sea-girt,
Elatreus, oar-shaft:
lover-of-the-sea, lover-of-the-
    sea-ebb,
lover-of-the-swift-sea,
Ponteus, Proreus, Oöos:
Anabesneos, who breaks to anger
as a wave to froth:
Amphiolos, one caught between
wave-shock and wave-shock:
Eurualos, board sea-wrack,
like Ares, man's death,
and Naubolidos, best in shape,
of all first in size:
Phaekous, sea's thunderbolt—
ah, crash on crash of great
    names—
man-tamer, man's-help, perfect
    Laodamos:
and last the sons of great
    Alkinöos,

Laodamos, Halios, and god-like
    Clytomeos.
Of all nations, of all cities,
of all continents,
she is favoured above the rest,
for she gives men as great as
    the sea,
to battle against the elements
    and evil:
greater even than the sea,
they live beyond wrack and
    death of cities,
and each god-like name spoken
is as a shrine in a godless place.
But to name you,
we, reverent, are breathless,
weak with pain and old loss,
and exile and despair—
our hearts break but to speak
your name, Oknaleos—
and may we but call you in the
    feverish wrack
of our storm-strewn beach,
    Eretmeos,
our hurt is quiet and our hearts
    tamed,
as the sea may yet be tamed,
and we vow to float great ships,
named for each hero,
and oar-blades, cut of mountain-
    trees
as such men might have shaped:
Eretmeos, and the sea is swept,
baffled by the lordly shape,

Akroneos has pines for his
    ship's keel;
to love, to mate the sea?
Ah there is Ponteos,
the very deep roar,
hailing you dear—
they clamour to Ponteos,
and to Proëos
leap, swift to kiss, to curl,
    to creep,
lover to mistress.
What wave, what love, what foam,
For Oöos who moves swift as
    the sea?
Ah stay, my heart, the weight
of lovers, of loneliness
drowns me,
alas that their very names
so press to break my heart
with heart-sick weariness,
what would they be,
the very gods,
rearing their mighty length
beside the unharvested sea?

'Thetis'

He had asked for immortal life
in the old days and had grown
     old,
now he had aged apace,
he asked for his youth
and I, Thetis, granted him
freedom under the sea
drip and welter of weeds,
the drift of the fringing grass,
the gift of the never-withering
     moss,
and the flowering reed,
and most,
beauty of fifty nereids,
sisters of nine,
I one of their least,
yet great and a goddess,
granted Pelius,
love under the sea,
beauty, grace infinite:
so I crept, at last,
a crescent, a curve of a wave,
(a man would have thought,
had he watched for his nets
on the beach)
a dolphin, a glistening fish,
that burnt and caught for its light,
the light of the undercrest
of the lifting tide,
a fish with silver for breast,
with no light but the light
of the sea it reflects.
Little he would have guessed,
(had such a one
watched by his nets,)
that a goddess flung from
     the crest
of the wave the blue of its own
bright tress of hair,
the blue of the painted stuff
it wore for dress.
No man would have known
     save he,
whose coming I sensed as
     I strung
my pearl and agate and pearl,
to mark the beat and the stress
of the lilt of my song.
*Who dreams of a son,
save one,
childless, having no bright
face to flatter its own,
who dreams of a son?
Nereids under the sea,
my sisters, fifty and one,
(counting myself)
they dream of a child
of water and sea,
with hair of the softest,
to lie along the curve
of fragile, tiny bones,
yet more beautiful each than each,
hair more bright and long,
to rival its own.
Nereids under the wave,
who dreams of a son
save I, Thetis, alone?
Each would have for a child,
a stray self, furtive and wild,
to dive and leap to the wind,
to wheedle and coax
the stray birds bright and bland
of foreign strands,
to crawl and stretch on the sands,
each would have for its own,
a daughter for child.
Who dreams, who sings of a son?
I, Thetis, alone.*
When I had finished my song,
and dropped the last seed-pearl,
and flung the necklet
about my throat

ABOVE: *The Return of Neptune,* painted by John Singleton Copley, c. 1754.
© Courtesy of The Met Museum, 59.198

and found it none too bright,
not bright enough nor pale
enough, not like the moon
    that creeps
beneath the sea,
between the lift of crest
    and crest,
had tried it on
and found it not
quite fair enough
to fill the night
of my blue folds of bluest dress
with moon for light,
I cast the beads aside and leapt,
myself all blue
with no bright gloss
of pearls for crescent light;
but one alert, all blue and wet,
I flung myself, an arrow's flight,
straight upward

through the blue of night
that was my palace wall,
and crept to where I saw the mark
of feet, a rare foot-fall:
Achilles' sandal on the beach,
could one mistake?
perhaps a lover or a nymph,
lost from the tangled fern
    and brake,
that lines the upper shelf of land,
perhaps a goddess or a nymph
might so mistake
Achilles' footprint for the trace
of a bright god alert to track
the panther where he slinks
    for thirst
across the sand;
perhaps a goddess or a nymph,
might think a god had crossed
    the track

of weed and drift,
had broken here this stem
    of reed,
had turned this sea shell to
    the light:
so she must stoop, this goddess
    girl,
or nymph, with crest of
    blossoming wood
about her hair for cap or crown,
must stoop and kneel and
    bending down,
must kiss the print of such a one
Not I, the mother, Thetis self,
I stretched and lay, a river's slim
dark length,
a rivulet where it leaves the wood,
and meets the sea,
I lay along the burning sand,
a river's blue.

184 'Sea-Heroes', 'Thetis' and 'At Ithaca' by H.D.

'At Ithaca'

Over and back,
the long waves crawl
and track the sand with foam;
night darkens and the sea
takes on that desperate tone
of dark that wives put on
when all their love is done.
Over and back,
the tangled thread falls slack,
over and up and on;
over and all is sewn;
now while I bind the end,
I wish some fiery friend
would sweep impetuously
these fingers from the loom.
My weary thoughts
play traitor to my soul,
just as the toil is over;
swift while the woof is whole,
turn now my spirit, swift,
and tear the pattern there,
the flowers so deftly wrought,
the borders of sea blue,
the sea-blue coast of home.
The web was over-fair,
that web of pictures there,
enchantments that I thought
he had, that I had lost;
weaving his happiness
within the stitching frame,
weaving his fire and frame,
I thought my work was done,
I prayed that only one
of those that I had spurned
might stoop and conquer this
long waiting with a kiss.
But each time that I see
my work so beautifully
inwoven and would keep
the picture and the whole,
Athene steels my soul.
Slanting across my brain,
I see as shafts of rain
his chariot and his shafts,
I see the arrows fall,
I see the lord who moves
like Hector lord of love,
I see him matched with fair
bright rivals, and I see
those lesser rivals flee.

ABOVE: Postcard of H.D., *c.* 1921.
© Courtesy of Wikimedia Commons

'DAGON' AND 'THE TEMPLE'

BY HP LOVECRAFT
(1919 AND 1925)

Howard Phillips Lovecraft (1890–1937) was born in Providence, Rhode Island. Throughout his life he produced short stories, novellas and poems, many set within a shared world we refer to as the Cthulhu Mythos, named for his most iconic monster. Lovecraft's contemporaries, including the writers Robert E Howard and Robert Bloch, used Lovecraft's settings and characters in their own work. Contemporary writers continue to do so today, reworking and adapting Lovecraft's creations in the form of films, role-playing games and digital games.

Lovecraft's stories often deal with the notion that there are things beyond human comprehension. His work is now often referred to as 'weird fiction' or 'cosmic horror' for its framing of human life as essentially meaningless and conviction of its eventual downfall. This nihilistic world view was linked to Lovecraft's xenophobia and fear that immigration would lead to the decline of his 'white, New England America', a recurring theme in many of his stories.

Lovecraft wrote four stories majorly featuring sea monsters: 'Dagon', 'The Temple', 'The Call of Cthulhu', and the novella *The Shadow Over Innsmouth*. The last two of these are omitted from this collection due to their length. 'Dagon' was first published in *The Vagrant* in 1919 then republished in *Weird Tales* in 1923. 'The Temple' was first published in *Weird Tales* in 1925. After Lovecraft's death at the age of 46, his contemporaries August Derleth and Donald Wandrei founded Arkham House to republish these stories in a collection titled *The Outsider and Others* (1939).

LEFT: The dust jacket from *The Outsider and Others*, illustrated by Virgil Finlay, 1939.
© Courtesy of Wikimedia Commons

'Dagon'

I am writing this under an appreciable mental strain, since by tonight I shall be no more. Penniless, and at the end of my supply of the drug which alone makes life endurable, I can bear the torture no longer; and shall cast myself from this garret window into the squalid street below. Do not think from my slavery to morphine that I am a weakling or a degenerate. When you have read these hastily scrawled pages you may guess, though never fully realise, why it is that I must have forgetfulness or death.

It was in one of the most open and least frequented parts of the broad Pacific that the packet of which I was supercargo fell a victim to the German sea-raider. The great war was then at its very beginning, and the ocean forces of the Hun had not completely sunk to their later degradation; so that our vessel was made a legitimate prize, whilst we of her crew were treated with all the fairness and consideration due us as naval prisoners. So liberal, indeed, was the discipline of our captors, that five days after we were taken I managed to escape alone in a small boat with water and provisions for a good length of time.

When I finally found myself adrift and free, I had but little idea of my surroundings. Never a competent navigator, I could only guess vaguely by the sun and stars that I was somewhat south of the equator. Of the longitude I knew nothing, and no island or coast-line was in sight. The weather kept fair, and for uncounted days I drifted aimlessly beneath the scorching sun; waiting either for some passing ship, or to be cast on the shores of some habitable land. But neither ship nor land appeared, and I began to despair in my solitude upon the heaving vastnesses of unbroken blue.

The change happened whilst I slept. Its details I shall never know; for my slumber, though troubled and dream-infested, was continuous. When at last I awaked, it was to discover myself half sucked into a slimy expanse of hellish black mire which extended about me in monotonous undulations as far as I could see, and in which my boat lay grounded some distance away.

Though one might well imagine that my first sensation would be of wonder at so prodigious and unexpected a transformation of scenery, I was in reality more horrified than astonished; for there was in the air and in the rotting soil a sinister quality which chilled me to the very core. The region was putrid with the carcasses of decaying fish, and of other less describable things which I saw protruding from the nasty mud of the unending plain. Perhaps I should not hope to convey in mere words the unutterable hideousness that can dwell in absolute silence and barren immensity. There was nothing within hearing, and nothing in sight save a vast reach of black slime; yet the very completeness of the stillness and the homogeneity of the landscape oppressed me with a nauseating fear.

The sun was blazing down from a sky which seemed to me almost black in its cloudless cruelty; as though reflecting the inky marsh beneath my feet. As I crawled into the stranded boat I realised that only one theory could explain my position. Through some unprecedented volcanic upheaval, a portion of the ocean floor must have been thrown to the surface, exposing regions which for innumerable millions of years had lain hidden under unfathomable watery depths. So great was the extent of the new land which had risen beneath me, that I could not detect the faintest noise of the surging ocean, strain my ears as I might. Nor were there any sea-fowl to prey upon the dead things.

For several hours I sat thinking or brooding in the boat, which lay upon its side and afforded a slight shade as the sun moved across the heavens. As the

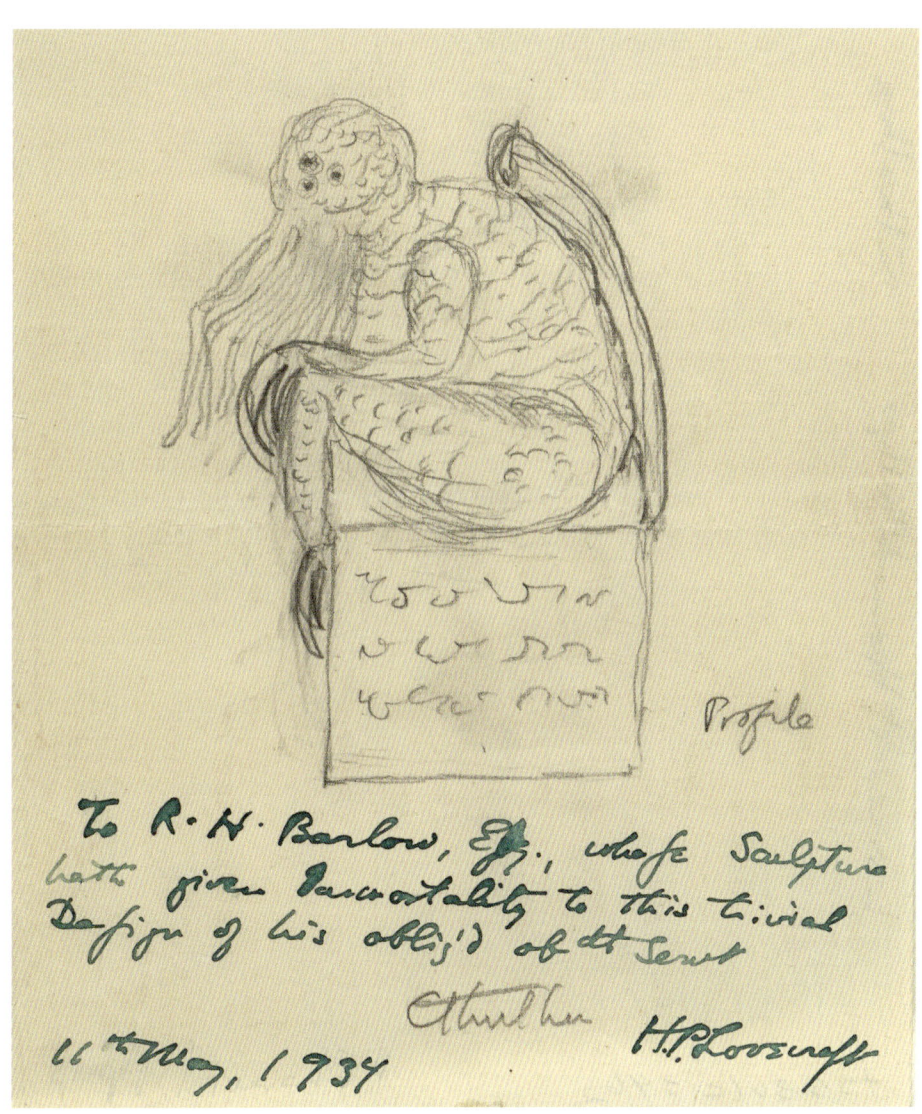

ABOVE: Illustration of Cthulhu by HP Lovecraft in a letter addressed to R. H. Barlow, dated 11 May 1934. The original copy is held at the Brown University Library Collections. © Courtesy of Wikimedia Commons.

'Dagon' and 'The Temple' by HP Lovecraft 189

day progressed, the ground lost some of its stickiness, and seemed likely to dry sufficiently for travelling purposes in a short time. That night I slept but little, and the next day I made for myself a pack containing food and water, preparatory to an overland journey in search of the vanished sea and possible rescue.

On the third morning I found the soil dry enough to walk upon with ease. The odour of the fish was maddening; but I was too much concerned with graver things to mind so slight an evil, and set out boldly for an unknown goal. All day I forged steadily westward, guided by a far-away hummock which rose higher than any other elevation on the rolling desert. That night I encamped, and on the following day still travelled toward the hummock, though that object seemed scarcely nearer than when I had first espied it. By the fourth evening I attained the base of the mound, which turned out to be much higher than it had appeared from a distance; an intervening valley setting it out in sharper relief from the general surface. Too weary to ascend, I slept in the shadow of the hill.

BELOW: Title illustration for the reprint of 'Dagon' in *Weird Tales*, October 1923, illustrated by William Heitman. 'Dagon' was first published in *The Vagrant* in 1919.
© Courtesy of Wikimedia Commons

I know not why my dreams were so wild that night; but ere the waning and fantastically gibbous moon had risen far above the eastern plain, I was awake in a cold perspiration, determined to sleep no more. Such visions as I had experienced were too much for me to endure again. And in the glow of the moon I saw how unwise I had been to travel by day. Without the glare of the parching sun, my journey would have cost me less energy; indeed, I now felt quite able to perform the ascent which had deterred me at sunset. Picking up my pack, I started for the crest of the eminence.

I have said that the unbroken monotony of the rolling plain was a source of vague horror to me; but I think my horror was greater when I gained the summit of the mound and looked down the other side into an immeasurable pit or canyon, whose black recesses the moon had not yet soared high enough to illumine. I felt myself on the edge of the world; peering over the rim into a fathomless chaos of eternal night. Through my terror ran curious reminiscences of Paradise Lost, and of Satan's hideous climb through the unfashioned realms of darkness.

As the moon climbed higher in the sky, I began to see that the slopes of the valley were not quite so perpendicular as I had imagined. Ledges and outcroppings of rock afforded fairly easy foot-holds for a descent, whilst after a drop of a few hundred feet, the declivity became very gradual. Urged on by an impulse which I cannot definitely analyse, I scrambled with difficulty down the rocks and stood on the gentler slope beneath, gazing into the Stygian deeps where no light had yet penetrated.

All at once my attention was captured by a vast and singular object on the opposite slope, which rose steeply about an hundred yards ahead of me; an object that gleamed whitely in the newly bestowed rays of the ascending moon. That it was merely a gigantic piece of stone, I soon assured myself; but I was conscious of a distinct impression that its contour and position were not altogether the work of Nature. A closer scrutiny filled me with sensations I cannot express; for despite its enormous magnitude, and its position in an abyss which had yawned at the bottom of the sea since the world was young, I perceived beyond a doubt that the strange object was a well-shaped monolith whose massive bulk had known the workmanship and perhaps the worship of living and thinking creatures.

Dazed and frightened, yet not without a certain thrill of the scientist's or archaeologist's delight, I examined my surroundings more closely. The moon, now near the zenith, shone weirdly and vividly above the towering steeps that hemmed in the chasm, and revealed the fact that a far-flung body of water flowed at the bottom, winding out of sight in both directions, and almost lapping my feet as I stood on the slope. Across the chasm, the wavelets washed the base of the Cyclopean monolith; on whose surface I could now trace both inscriptions and crude sculptures. The writing was in a system of hieroglyphics unknown to me, and unlike anything I had ever seen in books; consisting for the most part of conventionalised aquatic symbols such as fishes, eels, octopi, crustaceans, molluscs, whales, and the like. Several characters obviously represented marine things which are unknown to the modern world, but whose decomposing forms I had observed on the ocean-risen plain.

It was the pictorial carving, however, that did most to hold me spellbound. Plainly visible across the intervening water on account of their enormous size, were an array of bas-reliefs whose subjects would have excited the envy of a Doré. I think that these things were supposed to depict men—at least, a certain sort of men; though the creatures were shewn disporting like fishes in the waters of some marine grotto, or paying homage at some monolithic shrine which appeared to be under the waves as well. Of their faces and forms I dare not speak in detail; for the mere remembrance makes me grow faint. Grotesque beyond the imagination of a Poe or a Bulwer, they were damnably human in general outline despite webbed hands and feet, shockingly wide and flabby lips, glassy, bulging eyes, and other features less pleasant to recall. Curiously enough, they seemed to have been chiselled badly out of proportion with their scenic background; for one of the creatures was shewn in the act of killing a whale represented as but little larger than himself. I remarked, as I say, their grotesqueness and strange size; but in a moment decided that they were merely the imaginary gods of some primitive fishing or seafaring tribe; some tribe whose last descendant had perished eras before the first ancestor of the Piltdown or Neanderthal Man was born. Awestruck at this unexpected glimpse into a past beyond the conception of the most daring anthropologist, I stood musing whilst the moon cast queer reflections on the silent channel before me.

Then suddenly I saw it. With only a slight churning to mark its rise to the surface, the thing slid into view above the dark waters. Vast, Polyphemus-like, and loathsome, it darted like a stupendous monster of nightmares to the monolith, about which it flung its gigantic scaly arms, the while it bowed its hideous head and gave vent to certain measured sounds. I think I went mad then.

Of my frantic ascent of the slope and cliff, and of my delirious journey back to the stranded boat, I remember little. I believe I sang a great deal, and laughed oddly when I was unable to sing. I have indistinct recollections of a great storm some time after I reached the boat; at any rate, I know that I heard peals of thunder and other tones which Nature utters only in her wildest moods.

When I came out of the shadows I was in a San Francisco hospital; brought thither by the captain of the American ship which had picked up my boat in mid-ocean. In my delirium I had said much, but found that my words had been given scant attention. Of any land upheaval in the Pacific, my rescuers knew nothing; nor did I deem it necessary to insist upon a thing which I knew they could not believe. Once I sought out a celebrated ethnologist, and amused him with peculiar questions regarding the ancient Philistine legend of Dagon, the Fish-God; but soon perceiving that he was hopelessly conventional, I did not press my inquiries.

It is at night, especially when the moon is gibbous and waning, that I see the thing. I tried morphine; but the drug has given only transient surcease, and has drawn me into its clutches as a hopeless slave. So now I am to end it all, having written a full account for the information or the contemptuous amusement of my fellow-men. Often I ask myself if it could not all have been a pure phantasm—a mere freak of fever as I lay sun-stricken and raving in the open boat after my escape from the German man-of-war. This I ask myself, but ever does there come before me a hideously vivid vision in reply. I cannot think of the deep sea without shuddering at the nameless things that may at this very moment be crawling and floundering on its slimy bed, worshipping their ancient stone idols and carving their own detestable likenesses on submarine obelisks of water-soaked granite. I dream of a day when they may rise above the billows to drag down in their reeking talons the remnants of puny, war-exhausted mankind—of a day when the land shall sink, and the dark ocean floor shall ascend amidst universal pandemonium.

The end is near. I hear a noise at the door, as of some immense slippery body lumbering against it. It shall not find me. God, that hand! The window! The window!

'The Temple'

(Manuscript found on the coast of Yucatan.)

On August 20, 1917, I, Karl Heinrich, Graf von Altberg-Ehrenstein, Lieutenant-Commander in the Imperial German Navy and in charge of the submarine U-29, deposit this bottle and record in the Atlantic Ocean at a point to me unknown but probably about N. Latitude 20°, W. Longitude 35°, where my ship lies disabled on the ocean floor. I do so because of my desire to set certain unusual facts before the public; a thing I shall not in all probability survive to accomplish in person, since the circumstances surrounding me are as menacing as they are extraordinary, and involve not only the hopeless crippling of the U-29, but the impairment of my iron German will in a manner most disastrous.

On the afternoon of June 18, as reported by wireless to the U-61, bound for Kiel, we torpedoed the British freighter *Victory*, New York to Liverpool, in N. Latitude 45° 16′, W. Longitude 28° 34′; permitting the crew to leave in boats in order to obtain a good cinema view for the admiralty records. The ship sank quite picturesquely, bow first, the stern rising high out of the water whilst the hull shot down perpendicularly to the bottom of the sea. Our camera missed nothing, and I regret that so fine a reel of film should never reach Berlin. After that we sank the lifeboats with our guns and submerged.

When we rose to the surface about sunset a seaman's body was found on the deck, hands gripping the railing in curious fashion. The poor fellow was young, rather dark, and very handsome; probably an Italian or Greek, and undoubtedly of the *Victory*'s crew. He had evidently sought refuge on the very ship which had been forced to destroy his own—one more victim of the unjust war of aggression which the English pig-dogs are waging upon the Fatherland. Our men searched him for souvenirs, and found

in his coat pocket a very odd bit of ivory carved to represent a youth's head crowned with laurel. My fellow-officer, Lieut. Klenze, believed that the thing was of great age and artistic value, so took it from the men for himself. How it had ever come into the possession of a common sailor, neither he nor I could imagine.

As the dead man was thrown overboard there occurred two incidents which created much disturbance amongst the crew. The fellow's eyes had been closed; but in the dragging of his body to the rail they were jarred open, and many seemed to entertain a queer delusion that they gazed steadily and mockingly at Schmidt and Zimmer, who were bent over the corpse. The Boatswain Müller, an elderly man who would have known better had he not been a superstitious Alsatian swine, became so excited by this impression that he watched the body in the water; and swore that after it sank a little it drew its limbs into a swimming position and sped away to the south under the waves. Klenze and I did not like these displays of peasant ignorance, and severely reprimanded the men, particularly Müller.

The next day a very troublesome situation was created by the indisposition of some of the crew. They were evidently suffering from the nervous strain of our long voyage, and had had bad dreams. Several seemed quite dazed and stupid; and after satisfying myself that they were not feigning their weakness, I excused them from their duties. The sea was rather rough, so we descended to a depth where the waves were less troublesome. Here we were comparatively calm, despite a somewhat puzzling southward current which we could not identify from our oceanographic charts. The moans of the sick men were decidedly annoying; but since they did not appear to demoralise the rest of the crew, we did not resort to extreme measures. It was our plan to remain where we were and intercept the liner *Dacia*, mentioned in information from agents in New York.

In the early evening we rose to the surface, and found the sea less heavy. The smoke of a battleship was on the northern horizon, but our distance and ability to submerge made us safe. What worried us more was the talk of Boatswain Müller, which grew wilder as night came on. He was in a detestably childish state, and babbled of some illusion of dead bodies drifting past the undersea portholes; bodies which looked at him intensely, and which he recognised in spite of bloating as having seen dying during some of our victorious German exploits. And he said that the young man we had found and tossed overboard was their leader. This was very gruesome and abnormal, so we confined Müller in irons and had him soundly whipped. The men were not pleased at his punishment, but discipline was necessary. We also denied the request of a delegation headed by Seaman Zimmer, that the curious carved ivory head be cast into the sea.

On June 20, Seamen Bohm and Schmidt, who had been ill the day before, became violently insane. I regretted that no physician was included in our complement of officers, since German lives are precious; but the constant ravings of the two concerning a terrible curse were most subversive of discipline, so drastic steps were taken. The crew accepted the event in a sullen fashion, but it seemed to quiet Müller; who thereafter gave us no trouble. In the evening we released him, and he went about his duties silently.

In the week that followed we were all very nervous, watching for the *Dacia*. The tension was aggravated by the disappearance of Müller and Zimmer, who undoubtedly committed suicide as a result of the fears which had seemed to harass them, though they were not observed in the act of jumping overboard. I was rather glad to be rid of Müller, for even his silence had unfavourably affected the crew. Everyone seemed inclined to be silent now, as though holding a secret fear. Many were ill, but none made a disturbance. Lieut. Klenze chafed under the strain, and was annoyed by the merest trifles—such as the school of dolphins which gathered about the U-29 in increasing numbers, and the growing intensity of that southward current which was not on our chart.

It at length became apparent that we had missed the *Dacia* altogether. Such failures are not uncommon, and we were more pleased than disappointed; since our return to Wilhelmshaven was now in order. At noon June 28 we turned northeastward, and despite some rather comical entanglements with the unusual masses of dolphins were soon under way.

The explosion in the engine room at 2 P.M. was wholly a surprise. No defect in the machinery or carelessness in the men had been noticed, yet without warning the ship was racked from end to end with a colossal shock. Lieut. Klenze hurried to the engine room, finding the fuel-tank and most of the mechanism shattered, and Engineers Raabe and Schneider instantly killed. Our situation had suddenly become grave indeed; for though the chemical air regenerators were intact, and though we could use the devices for raising and submerging the ship and opening the hatches as long as compressed air and storage batteries might hold out, we were powerless to propel or guide the submarine. To seek rescue in the lifeboats would be to deliver ourselves into the hands of enemies unreasonably embittered against our great German nation, and our wireless had failed ever since the *Victory* affair to put us in touch with a fellow U-boat of the Imperial Navy.

From the hour of the accident till July 2 we drifted constantly to the south, almost without plans and encountering no vessel. Dolphins still encircled the U-29, a somewhat remarkable circumstance considering the distance we had covered. On the morning of July 2 we sighted a warship flying American colours, and the men became very restless in their desire to surrender. Finally Lieut. Klenze had to shoot a seaman named Traube, who urged this un-German act with especial violence. This quieted the crew for the time, and we submerged unseen.

The next afternoon a dense flock of sea-birds appeared from the south, and the ocean began to heave ominously. Closing our hatches, we awaited developments until we realised that we must either submerge or be swamped in the mounting waves. Our air pressure and electricity were diminishing, and we wished to avoid all unnecessary use of our slender mechanical resources; but in this case there was no choice. We did not descend far, and when after several hours the sea was calmer, we decided to return to the surface. Here, however, a new trouble developed; for the ship failed to respond to our direction in spite of all that the mechanics could do. As the men grew more frightened at this undersea imprisonment, some of them began to mutter again about Lieut. Klenze's ivory image, but the sight of an automatic pistol calmed them. We kept the poor devils as busy as we could, tinkering at the machinery even when we knew it was useless.

Klenze and I usually slept at different times; and it was during my sleep, about 5 A.M., July 4, that the general mutiny broke loose. The six remaining pigs of seamen, suspecting that we were lost, had suddenly burst into a mad fury at our refusal to surrender to the Yankee battleship two days before; and were in a delirium of cursing and destruction. They roared like the animals they were, and broke instruments and furniture indiscriminately; screaming about such nonsense as the curse of the ivory image and the dark dead youth who looked at them and swam away. Lieut. Klenze seemed paralysed and inefficient, as one might expect of a soft, womanish Rhinelander. I shot all six men, for it was necessary, and made sure that none remained alive.

We expelled the bodies through the double hatches and were alone in the U-29. Klenze seemed very nervous, and drank heavily. It was decided that we remain alive as long as possible, using the large stock of provisions and chemical supply of oxygen, none of which had suffered from the crazy antics of those swine-hound seamen. Our compasses, depth gauges, and other delicate instruments were ruined; so that henceforth our only reckoning would be guesswork, based on our watches, the calendar, and our apparent drift as judged by any objects we might spy through the portholes or from the conning tower. Fortunately we had storage batteries still capable of long use, both for interior lighting and for the searchlight. We often cast a beam around the ship, but saw only dolphins, swimming parallel to our own drifting course. I was scientifically interested in those dolphins; for though the ordinary Delphinus delphis is a cetacean mammal, unable to subsist without air, I watched one of the swimmers closely for two hours, and did not see him alter his submerged condition.

With the passage of time Klenze and I decided that we were still drifting south, meanwhile sinking deeper and deeper. We noted the marine fauna and flora, and read much on the subject in the books I had carried with me for spare moments. I could not help observing, however, the inferior scientific knowledge of my companion. His mind was not Prussian, but given to imaginings and speculations which have no value. The fact of our coming death affected him curiously, and he would frequently pray in remorse over the men, women, and children we had sent to the bottom; forgetting that all things are noble which serve the German state. After a time he became noticeably unbalanced, gazing for hours at his ivory image and weaving fanciful stories of the lost and forgotten things under the sea. Sometimes, as a psychological experiment, I would lead him on in these wanderings, and listen to his endless poetical quotations and tales of sunken ships. I was very sorry for him, for I dislike to see a German suffer; but he was not a good man to die with. For myself I was proud, knowing how the Fatherland would revere my memory and how my sons would be taught to be men like me.

On August 9, we espied the ocean floor, and sent a powerful beam from the searchlight over it. It was a vast undulating plain, mostly covered with seaweed, and strown with the shells of small molluscs. Here and there were slimy objects of puzzling contour, draped with weeds and encrusted with barnacles, which Klenze declared must be ancient ships lying in their graves. He was puzzled by one thing, a peak of solid matter, protruding above the ocean bed nearly four feet at its apex; about two feet thick, with flat sides and smooth upper surfaces which met at a very obtuse angle. I called the peak a bit of outcropping rock, but Klenze thought he saw carvings on it. After a while he began to shudder, and turned away from the scene as if frightened; yet could give no explanation save that he was overcome with the vastness, darkness, remoteness, antiquity, and mystery of the oceanic abysses. His mind was tired, but I am always a German, and was quick to notice two things; that the U-29 was standing the deep-sea pressure splendidly, and that the peculiar dolphins were still about us, even at a depth where the existence of high organisms is considered impossible by most naturalists. That I had previously overestimated our depth, I was sure; but none the less we must still be deep enough to make these phenomena remarkable. Our southward speed, as gauged by the ocean floor, was about as I had estimated from the organisms passed at higher levels.

It was at 3:15 P.M., August 12, that poor Klenze went wholly mad. He had been in the conning tower using the searchlight when I saw him bound into the library compartment where I sat reading, and his face at once betrayed him. I will repeat here what he said, underlining the words he emphasised: "He is calling! He is calling! I hear him! We must go!" As he spoke he took his ivory image from the table, pocketed it, and seized my arm in an effort to drag me up the companionway to the deck. In a moment I understood that he meant to open the hatch and plunge with me into the water outside, a vagary of suicidal and homicidal mania for which I was scarcely prepared. As I hung back and attempted to soothe him he grew more violent, saying: "Come now—do not wait until later; it is better to repent and be forgiven than to defy and be condemned." Then I tried the opposite of the soothing plan, and told him he was mad—pitifully demented. But he was unmoved, and cried: "If I am mad, it is mercy! May the gods pity the man who in his callousness can remain sane to the hideous end! Come and be mad whilst he still calls with mercy!"

This outburst seemed to relieve a pressure in his brain; for as he finished he grew much milder, asking me to let him depart alone if I would not accompany him. My course at once became clear. He was a German, but only a Rhinelander and a commoner; and he was now a potentially dangerous madman. By complying with his suicidal request I could immediately free myself from one who was no longer a companion but a menace. I asked him to give me the ivory image before he went, but this request brought from him such uncanny laughter that I did not repeat it. Then I asked him if he wished to leave any keepsake or lock of hair for his family in Germany in case I should be rescued, but again he gave me that strange laugh. So as he climbed the ladder I went to the levers, and allowing proper time-intervals operated the machinery which sent him to his death. After I saw that he was no longer in the boat I threw the searchlight around the water in an effort to obtain a last glimpse of him; since I wished to ascertain whether the water-pressure would flatten him as it theoretically should, or whether the body would be unaffected, like those extraordinary dolphins. I did not, however, succeed in finding my late companion, for the dolphins were massed thickly and obscuringly about the conning tower.

That evening I regretted that I had not taken the ivory image surreptitiously from poor Klenze's pocket as he left, for the memory of it fascinated me. I could not forget the youthful, beautiful head with its leafy crown, though I am not by nature an artist. I was also sorry that I had no one with whom to converse. Klenze, though not my mental equal, was much better than no one. I did not sleep well that night, and wondered exactly when the end would come. Surely, I had little enough chance of rescue.

The next day I ascended to the conning tower and commenced the customary searchlight explorations. Northward the view was much the same as it had been all the four days since we had sighted the bottom, but I perceived that the drifting of the U-29 was less rapid. As I swung the beam around to the south, I noticed that the ocean floor ahead fell away in a marked declivity, and bore curiously regular blocks of stone in certain places, disposed as if in accordance with definite patterns. The boat did not at once descend to match the greater ocean depth, so I was soon forced to adjust the searchlight to cast a sharply downward beam. Owing to the abruptness of the change a wire was disconnected, which necessitated a delay of many minutes for repairs; but at length the light streamed on again, flooding the marine valley below me.

I am not given to emotion of any kind, but my amazement was very great when I saw what lay revealed in that electrical glow. And yet as one reared in the best Kultur of Prussia I should not have been amazed, for geology and tradition alike tell us of great transpositions in oceanic and continental areas. What I saw was an extended and elaborate array of ruined edifices; all of magnificent though unclassified architecture, and in various stages of preservation. Most appeared to be of marble, gleaming whitely in the rays of the searchlight, and the general plan was of a large city at the bottom of a narrow valley, with numerous isolated temples and villas on the steep slopes above. Roofs were fallen and columns were broken, but there still remained an air of immemorially ancient splendour which nothing could efface.

Confronted at last with the Atlantis I had formerly

deemed largely a myth, I was the most eager of explorers. At the bottom of that valley a river once had flowed; for as I examined the scene more closely I beheld the remains of stone and marble bridges and sea-walls, and terraces and embankments once verdant and beautiful. In my enthusiasm I became nearly as idiotic and sentimental as poor Klenze, and was very tardy in noticing that the southward current had ceased at last, allowing the U-29 to settle slowly down upon the sunken city as an aëroplane settles upon a town of the upper earth. I was slow, too, in realising that the school of unusual dolphins had vanished.

In about two hours the boat rested in a paved plaza close to the rocky wall of the valley. On one side I could view the entire city as it sloped from the plaza down to the old river-bank; on the other side, in startling proximity, I was confronted by the richly ornate and perfectly preserved facade of a great building, evidently a temple, hollowed from the solid rock. Of the original workmanship of this titanic thing I can only make conjectures. The facade, of immense magnitude, apparently covers a continuous hollow recess; for its windows are many and widely distributed. In the centre yawns a great open door, reached by an impressive flight of steps, and surrounded by exquisite carvings like the figures of Bacchanals in relief. Foremost of all are the great columns and frieze, both decorated with sculptures of inexpressible beauty; obviously portraying idealised pastoral scenes and processions of priests and priestesses bearing strange ceremonial devices in adoration of a radiant god. The art is of the most phenomenal perfection, largely Hellenic in idea, yet strangely individual. It imparts an impression of terrible antiquity, as though it were the remotest rather than the immediate ancestor of Greek art. Nor can I doubt that every detail of this massive product was fashioned from the virgin hillside rock of our planet. It is palpably a part of the valley wall, though how the vast interior was ever excavated I cannot imagine. Perhaps a cavern or series of caverns furnished the nucleus. Neither age nor submersion has corroded the pristine grandeur of this awful fane—for fane indeed it must be—and today after thousands of years it rests untarnished and inviolate in the endless night and silence of an ocean chasm.

I cannot reckon the number of hours I spent in gazing at the sunken city with its buildings, arches, statues, and bridges, and the colossal temple with its beauty and mystery. Though I knew that death was near, my curiosity was consuming; and I threw the searchlight's beam about in eager quest. The shaft of light permitted me to learn many details, but refused to shew anything within the gaping door of the rock-hewn temple; and after a time I turned off the current, conscious of the need of conserving power. The rays were now perceptibly dimmer than they had been during the weeks of drifting. And as if sharpened by the coming deprivation of light, my desire to explore the watery secrets grew. I, a German, should be the first to tread those aeon-forgotten ways!

I produced and examined a deep-sea diving suit of joined metal, and experimented with the portable light and air regenerator. Though I should have trouble in managing the double hatches alone, I believed I could overcome all obstacles with my scientific skill and actually walk about the dead city in person.

On August 16 I effected an exit from the U-29, and laboriously made my way through the ruined and mud-choked streets to the ancient river. I found no skeletons or other human remains, but gleaned a wealth of archaeological lore from sculptures and coins. Of this I cannot now speak save to utter my awe at a culture in the full noon of glory when cave-dwellers roamed Europe and the Nile flowed unwatched to the sea. Others, guided by this manuscript if it shall ever be found, must unfold the mysteries at which I can only hint. I returned to the boat as my electric batteries grew feeble, resolved to explore the rock temple on the following day.

On the 17th, as my impulse to search out the mystery of the temple waxed still more insistent, a great disappointment befell me; for I found that the materials needed to replenish the portable light had perished in the mutiny of those pigs in July. My rage was unbounded, yet my German sense forbade me to venture unprepared into an utterly black interior which might prove the lair of some indescribable marine monster or a labyrinth of passages from whose windings I could never extricate myself. All I could do was to turn on the waning searchlight of the U-29, and with its aid walk up the temple steps and study the exterior carvings. The shaft of light entered the door at an upward angle, and I peered in to see if I could glimpse anything, but all in vain. Not even the roof was visible; and though I took a step or two inside after testing the floor with a staff, I dared not go farther. Moreover, for the first time in my life I experienced the emotion of dread. I began to realise how some of poor Klenze's moods had arisen, for as the temple drew me more and more, I feared its aqueous abysses with a blind and mounting terror. Returning to the submarine, I turned off the lights and sat thinking in the dark. Electricity must now be saved for emergencies.

Saturday the 18th I spent in total darkness, tormented by thoughts and memories that threatened to overcome my German will. Klenze had gone mad and perished before reaching this sinister remnant of a past unwholesomely remote, and had advised me to go with him. Was, indeed, Fate preserving my reason only to draw me irresistibly to an end more horrible and unthinkable than any man has dreamed of? Clearly, my nerves were sorely taxed, and I must cast off these impressions of weaker men.

I could not sleep Saturday night, and turned on the lights regardless of the future. It was annoying that the electricity should not last out the air and provisions. I revived my thoughts of euthanasia, and examined my automatic pistol. Toward morning I must have dropped asleep with the lights on, for I awoke in darkness yesterday afternoon to find the batteries dead. I struck several matches in succession, and desperately regretted the improvidence which had caused us long ago to use up the few candles we carried.

After the fading of the last match I dared to waste, I sat

RIGHT: Photograph of HP Lovecraft, taken in 1915. © Courtesy of Wikimedia Commons

very quietly without a light. As I considered the inevitable end my mind ran over preceding events, and developed a hitherto dormant impression which would have caused a weaker and more superstitious man to shudder. The head of the radiant god in the sculptures on the rock temple is the same as that carven bit of ivory which the dead sailor brought from the sea and which poor Klenze carried back into the sea.

I was a little dazed by this coincidence, but did not become terrified. It is only the inferior thinker who hastens to explain the singular and the complex by the primitive short cut of supernaturalism. The coincidence was strange, but I was too sound a reasoner to connect circumstances which admit of no logical connexion, or to associate in any uncanny fashion the disastrous events which had led from the *Victory* affair to my present plight. Feeling the need of more rest, I took a sedative and secured some more sleep. My nervous condition was reflected in my dreams, for I seemed to hear the cries of drowning persons, and to see dead faces pressing against the portholes of the boat. And among the dead faces was the living, mocking face of the youth with the ivory image.

I must be careful how I record my awaking today, for I am unstrung, and much hallucination is necessarily mixed with fact. Psychologically my case is most interesting, and I regret that it cannot be observed scientifically by a competent German authority. Upon opening my eyes my first sensation was an overmastering desire to visit the rock temple; a desire which grew every instant, yet which I automatically sought to resist through some emotion of fear which operated in the reverse direction. Next there came to me the impression of light amidst the darkness of dead batteries, and I seemed to see a sort of phosphorescent glow in the water through the porthole which opened toward the temple. This aroused my curiosity, for I knew of no deep-sea organism capable of emitting such luminosity. But before I could investigate there came a third impression which because of its irrationality caused me to doubt the objectivity of anything my senses might record. It was an aural delusion; a sensation of rhythmic, melodic sound as of some wild yet beautiful chant or choral hymn, coming from the outside through the absolutely sound-proof hull of the U-29. Convinced of my psychological and nervous abnormality, I lighted some matches and poured a stiff dose of sodium bromide solution, which seemed to calm me to the extent of dispelling the illusion of sound. But the phosphorescence remained, and I had difficulty in repressing a childish impulse to go to the porthole and seek its source. It was horribly realistic, and I could soon distinguish by its aid the familiar objects around me, as well as the empty sodium bromide glass of which I had had no former visual impression in its present location. The last circumstance made me ponder, and I crossed the room and touched the glass. It was indeed in the place where I had seemed to see it. Now I knew that the light was either real or part of an hallucination so fixed and consistent that I could not hope to dispel it, so abandoning all resistance I ascended to the conning tower to look for the luminous agency. Might it not actually be another

U-boat, offering possibilities of rescue?

It is well that the reader accept nothing which follows as objective truth, for since the events transcend natural law, they are necessarily the subjective and unreal creations of my overtaxed mind. When I attained the conning tower I found the sea in general far less luminous than I had expected. There was no animal or vegetable phosphorescence about, and the city that sloped down to the river was invisible in blackness. What I did see was not spectacular, not grotesque or terrifying, yet it removed my last vestige of trust in my consciousness. For the door and windows of the undersea temple hewn from the rocky hill were vividly aglow with a flickering radiance, as from a mighty altar-flame far within.

Later incidents are chaotic. As I stared at the uncannily lighted door and windows, I became subject to the most extravagant visions—visions so extravagant that I cannot even relate them. I fancied that I discerned objects in the temple—objects both stationary and moving—and seemed to hear again the unreal chant that had floated to me when first I awaked. And over all rose thoughts and fears which centred in the youth from the sea and the ivory image whose carving was duplicated on the frieze and columns of the temple before me. I thought of poor Klenze, and wondered where his body rested with the image he had carried back into the sea. He had warned me of something, and I had not heeded—but he was a soft-headed Rhinelander who went mad at troubles a Prussian could bear with ease.

The rest is very simple. My impulse to visit and enter the temple has now become an inexplicable and imperious command which ultimately cannot be denied. My own German will no longer controls my acts, and volition is henceforward possible only in minor matters. Such madness it was which drove Klenze to his death, bareheaded and unprotected in the ocean; but I am a Prussian and a man of sense, and will use to the last what little will I have. When first I saw that I must go, I prepared my diving suit, helmet, and air regenerator for instant donning; and immediately commenced to write this hurried chronicle in the hope that it may some day reach the world. I shall seal the manuscript in a bottle and entrust it to the sea as I leave the U-29 forever.

I have no fear, not even from the prophecies of the madman Klenze. What I have seen cannot be true, and I know that this madness of my own will at most lead only to suffocation when my air is gone. The light in the temple is a sheer delusion, and I shall die calmly, like a German, in the black and forgotten depths. This daemoniac laughter which I hear as I write comes only from my own weakening brain. So I will carefully don my diving suit and walk boldly up the steps into that primal shrine; that silent secret of unfathomed waters and uncounted years.

About five feet of hairy body was visible, and we perceived its eyes, which were as large as saucers, moving round slowly upon their long pedicles.

THE MARACOT DEEP

BY ARTHUR CONAN DOYLE
(1929)

Sir Arthur Conan Doyle (1859–1930), best known as the creator of Sherlock Holmes, was also a prolific author of early speculative fiction. He graduated from Edinburgh University in 1880 with an MBCM (Bachelor of Medicine and Master of Surgery) and practised as a ship's surgeon before opening an ophthalmology practice in London. Kicking off a period Sherlock Holmes fans refer to as 'The Great Hiatus', Conan Doyle killed off his titular character in a short story called 'The Final Problem' (1893) in order to focus on what he perceived to be more important literary work, including the novel *The Maracot Deep*, which was serialised in both *The Saturday Evening Post* and *The Strand Magazine* before publication in 1929.

LEFT: Illustration by Tom Peddie in *The Strand Magazine*, April–May 1929.
© Courtesy of The Arthur Conan Doyle Encyclopaedia

The Maracot Deep is the story of a team of marine scientists, led by the eccentric Professor Maracot to a trench in the Atlantic Ocean, the depth of which is thought to exceed that of the Challenger Deep in the Pacific Ocean. There is an impressive amount of scientific detail in Conan Doyle's description of the research vessel and the submersible, drawing upon the highly publicised Challenger Expedition of 1872–76 as well as Beebe and Barton's *Bathysphere*, designed in 1928–29 (see page 131). Professor Maracot's submersible is of scientifically plausible design, placing the story within the bounds of early science fiction, although the term was not yet in common usage.

As was the convention of the time, *The Maracot Deep* is extensively padded with a frame narrative, presenting the novel as a collection of letters and transcripts. This excerpt is taken from the first two chapters of the seven-chapter story.

The Maracot Deep

And then at last, quite softly and gently, we came to rest. So delicate was the impact that we should hardly have known of it had it not been that the light when turned on showed great coils of the hawser all around us. The wire was a danger to our breathing tubes, for it might foul them, and at the urgent cry of Maracot it was pulled taut from above once more. The dial marked eighteen hundred feet. We lay motionless on a volcanic ridge at the bottom of the Atlantic.

For a time I think that we all had the same feeling. We did not want to do anything or to see anything. We just wanted to sit quiet and try to realize the wonder of it—that we should be resting in the plumb centre of one of the great oceans of the world. But soon the strange scene round us, illuminated in all directions by our lights, drew us to the windows.

We had settled upon a bed of high algae ('*Cutleria multifida*,' said Maracot), the yellow fronds of which waved around us, moved by some deep-sea current, exactly as branches would move in a summer breeze. They were not long enough to obscure our view, though their great flat leaves, deep golden in the light, flowed occasionally across our vision. Beyond them lay slopes of some blackish slag-like material which were dotted with lovely coloured creatures, holothurians, ascidians, echini and echinoderms, as thickly as ever an English spring time bank was sprinkled with hyacinths and primroses. These living flowers of the sea, vivid scarlet, rich purple and delicate pink, were spread in profusion upon that coal-black background. Here and there great sponges bristled out from the crevices of the dark rocks, and a few fish of the middle depths, themselves showing up as flashes of colour, shot across our circle of vivid radiance. We were gazing enraptured at the fairy scene when an anxious voice came down the tube:

'Well, how do you like the bottom? Is all well with you? Don't be too long, for the glass is dropping and I don't like the look of it. Giving you air enough? Anything more we can do?'

'All right, Captain!' cried Maracot, cheerily. 'We won't be long. You are nursing us well. We are quite as comfortable as in our own cabin. Stand by presently to move us slowly forwards.'

We had come into the region of the luminous fishes, and it amused us to turn out our own lights, and in the absolute pitch-darkness—a darkness in which a sensitive plate can be suspended for an hour without a trace even of the ultra-violet ray—to look out at the phosphorescent activity of the ocean. As against a black velvet curtain one saw little points of brilliant light moving steadily along as a liner at night might shed light through its long line of portholes. One terrifying creature had luminous teeth which gnashed in Biblical fashion in the outer darkness. Another had long golden antennae, and yet another a plume of flame above its head.

As far as our vision carried, brilliant points flashed in the darkness, each little being bent upon its own business, and lighting up its own course as surely as the nightly taxicab at the theatre-hour in the Strand. Soon we had our own lights up again and the Doctor was making his observations of the sea-bottom.

'Deep as we are, we are not deep enough to get any of the characteristic Bathic deposits,' said he. 'These are entirely beyond our possible range. Perhaps on another occasion with a longer hawser–'

'Cut it out!' growled Bill. 'Forget it!'

Maracot smiled. 'You will soon get acclimatized to the depths, Scanlan. This will not be our only descent.'

'The Hell you say!' muttered Bill.

'You will think no more of it than of going down into the hold of the Stratford. You will observe, Mr. Headley, that the groundwork here, so far as we can observe it through the dense growth of hydrozoa and silicious sponges, is pumicestone and the black slag of basalt, pointing to ancient plutonic activities. Indeed, I am inclined to think that it confirms my previous view that this ridge is part of a volcanic formation and that the Maracot Deep,' he rolled out the words as if he loved them, 'represents the outer slope of the mountain. It has struck me that it would be an interesting experiment to move our cage slowly onwards until we come to the edge of the Deep, and see exactly what the formation may be at that point. I should expect to find a precipice of majestic dimensions extending downwards at a sharp angle into the extreme depths of the ocean.'

The experiment seemed to me to be a dangerous one, for who could say how far our thin hawser could bear the strain of lateral movement; but with Maracot danger, either to himself or to anyone else, simply did not exist when a scientific observation had to be made. I held my breath, and so I observed did Bill Scanlan, when a slow movement of our steel shell, brushing aside the waving fronds of seaweed, showed that the full strain was upon the line. It stood it nobly, however, and with a very gentle sweeping progression we began to glide over the bottom of the ocean, Maracot, with a compass in the hollow of his hand, shouting his direction as to the course to follow, and occasionally ordering the shell to be raised so as to avoid some obstacle in our path.

'This basaltic ridge can hardly be more than a mile across,' he explained. 'I had marked the abyss as being to the west of the point where we took our plunge. At this rate, we should certainly reach it in a very short time.'

We slid without any check over the volcanic plain, all feathered by the waving golden algae and made beautiful by the gorgeous jewels of Nature's cutting, flaming out from their setting of jet. Suddenly the Doctor dashed to the telephone.

'Stop her!' he cried. 'We are there!'

A monstrous gap had opened suddenly before us. It was a fearsome place, the vision of a nightmare. Black shining cliffs of basalt fell sheer down into the unknown. Their edges were fringed with dangling laminaria as ferns might overhang some earthly gorge, but beneath that tossing, vibrating rim there were only the black gleaming walls of the chasm. The rocky edge curved away from us, but the abyss might be of any breadth,

ABOVE: An ad for *The Maracot Deep*, in *The Bridgeport Telegram*, 6 October 1927.
© Courtesy of The Arthur Conan Doyle Encyclopaedia

for our lights failed to penetrate the gloom which lay before us. When a Lucas signalling lamp was turned downwards it shot out a long golden lane of parallel beams extending down, down, down until it was quenched in the gloom of the terrible chasm beneath us.

'It is indeed wonderful!' cried Maracot, gazing out with a pleased proprietary expression upon his thin, eager face. 'For depth I need not say that it has often been exceeded. There is the Challenger Deep of twenty-six thousand feet near the Ladrone Islands, the Planet Deep of thirty-two thousand feet off the Philippines, and many others, but it is probable that the Maracot Deep stands alone in the declivity of its descent, and is remarkable also for its escape from the observation of so many hydrographic explorers who have charted the Atlantic. It can hardly be doubted–'

He had stopped in the middle of a sentence and a look of intense interest and surprise had frozen upon his face. Bill Scanlan and I, gazing over his shoulders, were petrified by that which met our startled eyes.

Some great creature was coming up the tunnel of light

which we had projected into the abyss. Far down where it tailed off into the darkness of the pit we could dimly see the vague black lurchings and heavings of some monstrous body in slow upward progression. Paddling in clumsy fashion, it was rising with dim flickerings to the edge of the gulf. Now, as it came nearer, it was right in the beam, and we could see its dreadful form more clearly. It was a beast unknown to Science, and yet with an analogy to much with which we are familiar. Too long for a huge crab and too short for a giant lobster, it was moulded more upon the lines of the crayfish, with two monstrous nippers outstretched on either side, and a pair of sixteen-foot antennae which quivered in front of its black dull sullen eyes. The carapace, light yellow in colour, may have been ten feet across, and its total length, apart from the antennae, must have been not less than thirty.

'Wonderful!' cried Maracot, scribbling desperately in his notebook. 'Semi-pediculated eyes, elastic lamellae, family crustacea, species unknown. *Crustaceus Maracoti*—why not? Why not?'

'By gosh, I'll pass its name, but it seems to me it's coming our way!' cried Bill. 'Say, Doc, what about putting our light out?'

'Just one moment while I note the reticulations!' cried the naturalist. 'Yes, yes, that will do.' He clicked off the switch and we were back in our inky darkness, with only the darting lights outside like meteors on a moonless night.

'That beast is sure the world's worst,' said Bill, wiping his forehead. 'I felt like the morning after a bottle of Prohibition Hoosh.'

'It is certainly terrible to look at,' Maracot remarked, 'and perhaps terrible to deal with also if we were really exposed to those monstrous claws. But inside our steel case we can afford to examine him in safety and at our ease.'

He had hardly spoken when there came a rap as from a pickaxe upon our outer wall. Then there was a long drawn rasping and scratching, ending in another sharp rap.

'Say, he wants to come in!' cried Bill Scanlan in alarm. 'By gosh! we want "No Admission" painted on this shack.' His shaking voice showed how forced was his merriment, and I confess that my own knees were knocking together as I was aware of the stealthy monster closing up with an even blacker darkness each of our windows in succession, as he explored this strange shell which, could he but crack it, might contain his food.

'He can't hurt us,' said Maracot, but there was less assurance in his tone. 'Maybe it would be as well to shake the brute off.' He hailed the Captain up the tube.

'Pull us up twenty or thirty feet,' he cried.

A few seconds later we rose from the lava plain and swung gently in the still water. But the terrible beast was pertinacious. After a very short interval we heard once more the raspings of his feelers and the sharp tappings of his claws as he felt us round. It was terrible to sit silently in the dark and know that death was so near. If that mighty claw fell upon the window, would it stand the strain? That was the unspoken question in each of our minds.

But suddenly an unexpected and more urgent danger presented itself. The tappings had gone to the roof of our little dwelling, and now we began to sway with a rhythmic movement to and fro.

'Good God!' I cried. 'It has hold of the hawser. It will surely snap it.'

'Say, Doc, it's mine for the surface. I guess we've seen what we came to see, and it's home, sweet home for Bill Scanlan. Ring up the elevator and get her going.'

'But our work is not half done,' croaked Maracot. 'We have only begun to explore the edges of the Deep. Let us at least see how broad it is. When we have reached the other side I shall be content to return.' Then up the tube: 'All well, Captain. Move on at two knots until I call for a stop.'

We moved slowly out over the edge of the abyss. Since darkness had not saved us from attack we now turned on our lights. One of the portholes was entirely obscured by what appeared to be the creature's lower stomach. Its head and its great nippers were at work above us, and we still swayed like a clanging bell. The strength of the beast must have been enormous. Were ever mortals placed in such a situation, with five miles of water beneath— and that deadly monster above? The oscillations became more and more violent. An excited shout came down the tube from the Captain as he became aware of the jerks upon the hawser, and Maracot sprang to his feet with his hands thrown upwards in despair. Even within the shell we were aware of the jar of the broken wires, and an instant later we were falling into the mighty gulf beneath us.

As I look back at that awful moment I can remember hearing a wild cry from Maracot. 'The hawser has parted! You can do nothing! We are all dead men!' he yelled, grabbing at the telephone tube, and then, 'Good-bye, Captain, good-bye to all.' They were our last words to the world of men.

We did not fall swiftly down, as you might have imagined. In spite of our weight our hollow shell gave us some sustaining buoyancy, and we sank slowly and gently into the abyss. I heard the long scrape as we slid through the claws of the horrible creature who had been our ruin, and then with a smooth gyration we went circling downwards into the abysmal depths. It may have been fully five minutes, and it seemed like an hour, before we reached the limit of our telephone wire and snapped it like a thread. Our air tube broke off at almost the same moment and the salt water came spouting through the vents. With quick, deft hands Bill Scanlan tied cords round each of the rubber tubes and so stopped the inrush, while the Doctor released the top of our compressed air which came hissing forth from the tubes. The lights had gone out when the wire snapped, but even in the dark the Doctor was able to connect up the Hellesens dry cells which lit a number of lamps in the roof.

'It should last us a week,' he said, with a wry smile. 'We shall at least have light to die in.' Then he shook his head sadly and a kindly smile came over his gaunt features. 'It is all right for me. I am an old man and have done my work in the world. My one regret is that I should have allowed you two young fellows to come with me. I should have taken the risk alone.'

I simply shook his hand in reassurance, for indeed there was nothing I could say. Bill Scanlan, too, was silent. Slowly we sank, marking our pace by the dark fish shadows which flitted past our windows. It seemed as if they were flying upwards rather than that we were sinking down. We still oscillated, and there was nothing so far as I could see to prevent us from falling on our side, or even turning upside down. Our weight, however, was, fortunately, very evenly balanced and we kept a level floor. Glancing up at the bathymeter I saw that we had already reached the depth of a mile.

'You see, it is as I said,' remarked Maracot, with some complacency. 'You may have seen my paper in the Proceedings of the Oceanographical Society upon the relation of pressure and depth. I wish I could get one word back to the world, if only to confute Bulow of Giessen, who ventured to contradict me.'

'My gosh! If I could get a word back to the world I wouldn't waste it on a square-head highbrow,' said the mechanic. 'There is a little wren in Philadelphia that will have tears in her pretty eyes when she hears that Bill Scanlan has passed out. Well, it sure does seem a darned queer way of doing it, anyhow.'

'You should never have come,' I said, putting my hand on his.

'What sort of tin-horn sport should I have been if I had quitted?' he answered. 'No, it's my job, and I am glad I stuck it.'

'How long have we?' I asked the Doctor, after a pause.

He shrugged his shoulders.

'We shall have time to see the real bottom of the ocean, anyhow,' said he. 'There is air enough in our tubes for the best part of a day. Our trouble is with the waste products. That is what is going to choke us. If we could get rid of our carbon dioxide–'

'That I can see is impossible.'

'There is one tube of pure oxygen. I put it in in case of accidents. A little of that from time to time will help to keep us alive. You will observe that we are now more than two miles deep.'

'Why should we try to keep ourselves alive? The sooner it is over the better,' said I.

'That's the dope,' cried Scanlan. 'Cut loose and have done with it.'

'And miss the most wonderful sight that man's eye has ever seen!' said Maracot. 'It would be treason to Science. Let us record facts to the end, even if they should be for ever buried with our bodies. Play the game out.'

'Some sport, the Doc!' cried Scanlan. 'I guess he has the best guts of the bunch. Let us see the spiel to an end.'

We sat patiently on the settee, the three of us, gripping the edges of it with strained fingers as it swayed and rocked, while the fishes still flashed swiftly upwards athwart the portholes.

'It is now three miles,' remarked Maracot. 'I will turn on the oxygen, Mr. Headley, for it is certainly very close. There is one thing,' he added, with his dry, cackling laugh, 'it will certainly be the Maracot Deep from this time onwards. When Captain Howie takes back the news my colleagues will see to it that my grave is also my monument. Even Bulow of Giessen–' He babbled on about some unintelligible scientific grievance.

We sat in silence again, watching the needle as it crawled on to its fourth mile. At one point we struck something heavy, which shook us so violently that I feared that we would turn upon our side. It may have been a huge fish, or conceivably we may have bumped upon some projection of the cliff over the edge of which we had been precipitated. That edge had seemed to us at the time to be such a wondrous depth, and now looking back at it from our dreadful abyss it might almost have been the surface. Still we swirled and circled lower and lower through the dark green waste of waters. Twenty-five thousand feet now was registered upon the dial.

'We are nearly at our journey's end,' said Maracot. 'My Scott's recorder gave me twenty-six thousand seven hundred last year at the deepest point. We shall know our fate within a few minutes. It may be that the shock will crush us. It may be—'

And at that moment we landed.

There was never a babe lowered by its mother on to a feather-bed who nestled down more gently than we on to the extreme bottom of the Atlantic Ocean. The soft thick elastic ooze upon which we lit was a perfect buffer, which saved us from the slightest jar. We hardly moved upon our seats, and it is as well that we did not, for we had perched upon some sort of a projecting hummock, clothed thickly with the viscous gelatinous mud, and there we were balanced rocking gently with nearly half our base projecting and unsupported. There was a danger that we would tip over on our side, but finally we steadied down and remained motionless. As we did so Dr. Maracot, staring out through his porthole, gave a cry of surprise and hurriedly turned out our electric light.

To our amazement we could still see clearly. There was a dim, misty light outside which streamed through our porthole, like the cold radiance of a winter morning. We looked out at the strange scene, and with no help from our own lights we could see clearly for some hundred yards in each direction. It was impossible, inconceivable, but none the less the evidence of our senses told us that it was a fact. The great ocean floor is luminous.

'Why not?' cried Maracot, when we had stood for a minute or two in silent wonder. 'Should I not have foreseen it? What is this pteropod or globigerina ooze? Is it not the product of decay, the mouldering bodies of a billion billion organic creatures? And is decay not associated with phosphorescent luminosity? Where, in all creation, would it be seen if it were not here? Ah! It is indeed hard that we should have such a demonstration and be unable to send our knowledge back to the world.'

'And yet,' I remarked, 'we have scooped half a ton of radiolarian jelly at a time and detected no such radiance.'

'It would lose it, doubtless, in its long journey to the surface. And what is half a ton compared to these far-stretching plains of slow putrescence? And see, see,' he cried in uncontrollable excitement, 'the deep-sea creatures graze upon this organic carpet even as our herds on land graze upon the meadows!'

As he spoke a flock of big black fish, heavy and squat, came slowly over the ocean bed towards us, nuzzling among the spongy growths and nibbling away as they advanced. Another huge red creature, like a foolish cow of the ocean, was chewing the cud in front of my porthole, and others were grazing here and there, raising their heads from, time to time to gaze at this strange object which had so suddenly appeared among them.

I could only marvel at Maracot, who in that foul atmosphere, seated under the very shadow of death, still obeyed the call of Science and scribbled his observations in his notebook. Without following his precise methods, I none

About five feet of hairy body was visible, and we perceived its eyes, which were as large as saucers, moving round slowly upon their long pedicles.

ABOVE: Illustration by Tom Peddie in *The Strand Magazine,* April–May 1929.
© Courtesy of The Arthur Conan Doyle Encyclopaedia

the less made my own mental notes, which will remain for ever as a picture stamped upon my brain. The lowest plains of ocean consist of red clay, but here it was overlaid by the grey bathybian slime which formed an undulating plain as far as our eyes could reach. This plain was not smooth, but was broken by numerous strange rounded hillocks like that upon which we had perched, all glimmering in the spectral light. Between these little hills there darted great clouds of strange fish, many of them quite unknown to Science, exhibiting every shade of colour, but black and red predominating. Maracot watched them with suppressed excitement and chronicled them in his notes.

The air had become very foul, and again we were only able to save ourselves by a fresh emission of oxygen. Curiously enough, we were all hungry—I should rather say ravenous—and we fell upon the potted beef with bread and butter, washed down by whisky and water, which the foresight of Maracot had provided. With my perceptions stimulated by this refreshment, I was seated at my lookout portal and longing for a last cigarette, when my eyes caught something which sent a whirl of strange thoughts and anticipations through my mind.

I have said that the undulating grey plain on every side of us was studded with what seemed like hummocks. A particularly large one was in front of my porthole, and I looked out at it within a range

The Maracot Deep by Arthur Conan Doyle 213

of thirty feet. There was some peculiar mark upon the side of it, and as I glanced along I saw to my surprise that this mark was repeated again and again until it was lost round the curve. When one is so near death it takes much to give one a thrill about anything connected with this world, but my breath failed me for a moment and my heart stood still as I suddenly realized that it was a frieze at which I was looking and that, barnacled and worn as it was, the hand of man had surely at some time carved these faded figures. Maracot and Scanlan crowded to my porthole and gazed out in utter amazement at these signs of the omnipresent energies of man.

'It is carving, for sure!' cried Scanlan. 'I guess this dump has been the roof of a building. Then these other ones are buildings also. Say, boss, we've dropped plumb on to a regular burg.'

'It is, indeed, an ancient city,' said Maracot. 'Geology teaches that the seas have once been continents and the continents seas, but I have always distrusted the idea that in times so recent as the quaternary there could have been an Atlantic subsidence. Plato's report of Egyptian gossip had then a foundation of fact. These volcanic formations confirm the view that this subsidence was due to seismic activity.'

'There is regularity about these domes,' I remarked. 'I begin to think that they are not separate houses, but that they are cupolas and form the ornaments of the roof of some huge building.'

'I guess you are right,' said Scanlan. 'There are four big ones at the corners and the small ones in lines between. It's some building, if we could see the whole of it! You could put the whole Merribank plant inside it—and then some.'

'It has been buried up to the roof by the constant dropping from above,' said Maracot. 'On the other hand, it has not decayed. We have a constant temperature of a little over 32 Fahrenheit in the great depths, which would arrest destructive processes. Even the dissolution of the Bathic remains which pave the floor of the ocean and incidentally give us this luminosity must be a very slow one. But, dear me! this marking is not a frieze but an inscription.'

There was no doubt that he was right. The same symbol recurred every here and there. These marks were unquestionably letters of some archaic alphabet.

'I have made a study of Phoenician antiquities, and there is certainly something suggestive and familiar in these characters,' said our leader. 'Well, we have seen a buried city of ancient days, my friends, and we carry a wonderful piece of knowledge with us to the grave. There is no more to be learned. Our book of knowledge is closed. I agree with you that the sooner the end comes the better.'

It could not now be long delayed. The air was stagnant and dreadful. So heavy was it with carbon products that the oxygen could hardly force its way out against the pressure. By standing on the settee one was able to get a gulp of purer air, but the mephitic reek was slowly rising. Dr. Maracot folded his arms with an air of resignation and sank his head upon his breast. Scanlan was now overpowered by the fumes and was already sprawling upon the floor. My own head was swimming, and I felt an intolerable weight at my chest. I closed my eyes and my senses were rapidly slipping away.

Then I opened them for one last glimpse of that world which I was leaving, and as I did so I staggered to my feet with a hoarse scream of amazement.

A human face was looking in at us through the porthole!

Was it my delirium? I clutched at the shoulder of Maracot and shook him violently. He sat up and stared, wonder-struck and speechless at this apparition. If he saw it as well as I, it was no figment of the brain. The face was long and thin, dark in complexion, with a short, pointed beard, and two vivid eyes darting here and there in quick, questioning glances which took in every detail of our situation. The utmost amazement was visible upon the man's face. Our lights were now full on, and it must indeed have been a strange and vivid picture which presented itself to his gaze in that tiny chamber of death, where one man lay senseless and two others glared out at him with the twisted, contorted features of dying men, cyanosed by incipient asphyxiation. We both had our hands to our throats, and our heaving chests carried their message of despair. The man gave a wave of his hand and hurried away.

ABOVE: Spirit photograph of Arthur Conan Doyle, taken by Ada Adam Deane, *c.* 1922.
© Courtesy of Wikimedia Commons

'He has deserted us!' cried Maracot.

'Or gone for help. Let us get Scanlan on the couch. It's death for him down there.'

We dragged the mechanic on to the settee and propped his head against the cushions. His face was grey and he murmured in delirium, but his pulse was still perceptible.

'There is hope for us yet,' I croaked.

'But it is madness!' cried

The Maracot Deep by Arthur Conan Doyle

Maracot. 'How can man live at the bottom of the ocean? How can he breathe? It is collective hallucination. My young friend, we are going mad.'

Looking out at the bleak, lonely grey landscape in the dreary spectral light, I felt that it might be as Maracot said. Then suddenly I was aware of movement. Shadows were flitting through the distant water. They hardened and thickened into moving figures. A crowd of people were hurrying across the ocean bed in our direction. An instant later they had assembled in front of the porthole and were pointing and gesticulating in animated debate. There were several women in the crowd, but the greater part were men, one of whom, a powerful figure with a very large head and a full black beard, was clearly a person of authority. He made a swift inspection of our steel shell, and, since the edge of our base projected over the place on which we rested, he was able to see that there was a hinged trap-door at the bottom. He now sent a messenger flying back, while he made energetic and commanding signs to us to open the door from within.

'Why not?' I asked. 'We may as well be drowned as be smothered. I can stand it no longer.'

'We may not be drowned,' said Maracot. 'The water entering from below cannot rise above the level of the compressed air. Give Scanlan some brandy. He must make an effort, if it is his last one.'

I forced a drink down the mechanic's throat. He gulped and looked round him with wondering eyes. Between us we got him erect on the settee and stood on either side of him. He was still half-dazed, but in a few words I explained the situation.

'There is a chance of chlorine poisoning if the water reaches the batteries,' said Maracot. 'Open every air tube, for the more pressure we can get the less water may enter. Now help me while I pull upon the lever.'

We bent our weight upon it and yanked up the circular plate from the bottom of our little home, though I felt like a suicide as I did so. The green water, sparkling and gleaming under our light, came gurgling and surging in. It rose rapidly to our feet, to our knees, to our waists, and there it stopped. But the pressure of the air was intolerable. Our heads buzzed and the drums of our ears were bursting. We could not have lived in such an atmosphere for long. Only by clutching at the rack could we save ourselves from falling back into the waters beneath us.

From our higher position we could no longer see through the portholes, nor could we imagine what steps were being taken for our deliverance. Indeed, that any effective help could come to us seemed beyond the power of thought, and yet there was a commanding and purposeful air about these people, and especially about that squat bearded chieftain, which inspired vague hopes. Suddenly we were aware of his face looking up at us through the water beneath and an instant later he had passed through the circular opening and had clambered on to the settee, so that he was standing by our side—a short sturdy figure, not higher than my shoulder, but surveying us with large brown eyes, which were full of a half-amused confidence, as who should say, 'You poor devils; you think you are in a very bad way, but I can clearly see the road out.'

Only now was I aware of a very amazing thing. The man,

if indeed he was of the same humanity as ourselves, had a transparent envelope all round him which enveloped his head and body, while his arms and legs were free. So translucent was it that no one could detect it in the water, but now that he was in the air beside us it glistened like silver, though it remained as clear as the finest glass. On either shoulder he had a curious rounded projection beneath the clear protective sheath. It looked like an oblong box pierced with many holes, and gave him an appearance as if he were wearing epaulettes.

When our new friend had joined us another face appeared in the aperture of the bottom and thrust through it what seemed like a great bubble of glass. Three of these in succession were passed in and floated upon the surface of the water. Then six small boxes were handed up and our new acquaintance tied one with the straps attached to them to each of our shoulders, whence they stood up like his own. Already I began to surmise that no infraction of natural law was involved in the life of these strange people, and that while one box in some new fashion was a producer of air the other was an absorber of waste products. He now passed the transparent suits over our heads, and we felt that they clasped us tightly in the upper arm and waist by elastic bands, so that no water could penetrate. Within we breathed with perfect ease, and it was a joy to me to see Maracot looking out at me with his eyes twinkling as of old behind his glasses, while Bill Scanlan's grin assured me that the life-giving oxygen had done its work, and that he was his cheerful self once more. Our rescuer looked from one to another of us with grave satisfaction, and then motioned to us to follow him through the trap-door and out on to the floor of the ocean. A dozen willing hands were outstretched to pull us through and to sustain our first faltering steps as we staggered with our feet deep in the slimy ooze.

Even now I cannot get past the marvel of it! There we were, the three of us, unhurt and at our ease at the bottom of a five-mile abyss of water. Where was that terrific pressure which had exercised the imagination of so many scientists? We were no more affected by it than were the dainty fish which swam around us. It is true that, so far as our bodies were concerned, we were protected by these delicate bells of vitrine, which were in truth tougher than the strongest steel, but even our limbs, which were exposed, felt no more than a firm constriction from the water which one learned in time to disregard. It was wonderful to stand together and to look back at the shell from which we had emerged. We had left the batteries at work, and it was a wondrous object with its streams of yellow light flooding out from each side, while clouds of fishes gathered at each window. As we watched it the leader took Maracot by the hand, and we followed them both across the watery morass, clumping heavily through the sticky surface.

And now a most surprising incident occurred, which was clearly as astonishing to these strange new companions of ours as to ourselves. Above our heads there appeared a small, dark object, descending from the darkness above us and swinging down until it reached the bed of the ocean within a very short distance from where

The Maracot Deep by Arthur Conan Doyle

we stood. It was, of course, the deep-sea lead from the Stratford above us, making a sounding of that watery gulf with which the name of the expedition was to be associated. We had seen it already upon its downward path, and we could well understand that the tragedy of our disappearance had suspended the operation, but that after a pause it had been concluded, with little thought that it would finish almost at our feet. They were unconscious, apparently, that they had touched bottom, for the lead lay motionless in the ooze. Above me stretched the taut piano wire which connected me through five miles of water with the deck of our vessel. Oh, that it were possible to write a note and to attach it! The idea was absurd, and yet could I not send some message which would show them that we were still conscious? My coat was covered by my glass bell and the pockets were unapproachable, but I was free below the waist and my handkerchief chanced to be in my trousers pocket. I pulled it out and tied it above the top of the lead. The weight at once disengaged itself by its automatic mechanism, and presently I saw my white wisp of linen flying upwards to that world which I may never see again. Our new acquaintances examined the seventy-five pounds of lead with great interest, and finally carried it off with us as we went upon our way.

We had only walked a couple of hundred yards, threading our way among the hummocks, when we halted before a small square-cut door with solid pillars on either side and an inscription across the lintel. It was open, and we passed through it into a large, bare chamber. There was a sliding partition worked by a crank from within, and this was drawn across behind us. We could, of course, hear nothing in our glass helmets, but after standing a few minutes we were aware that a powerful pump must be at work, for we saw the level of the water sinking rapidly above us. In less than a quarter of an hour we were standing upon a sloppy stone-flagged pavement, while our new friends were busy in undoing the fastenings of our transparent suits. An instant later there we stood, breathing perfectly pure air in a warm, well-lighted atmosphere, while the dark people of the abyss, smiling and chattering, crowded round us with hand-shakings and friendly pattings. It was a strange, rasping tongue that they spoke, and no word of it was intelligible to us, but the smile on the face and the light of friendship in the eye are understandable even in the waters under the earth. The glass suits were hung on numbered pegs upon the wall, and the kindly folk half led and half pushed us to an inner

door which opened on to a long downward-sloping corridor. When it closed again behind us there was nothing to remind us of the stupendous fact that we were the involuntary guests of an unknown race at the bottom of the Atlantic ocean and cut off for ever from the world to which we belonged.

Now that the terrific strain had been so suddenly eased we were all exhausted. Even Bill Scanlan, who was a pocket Hercules, dragged his feet along the floor, while Maracot and I were only too glad to lean heavily upon our guides. Yet, weary as I was, I took in every detail as we passed. That the air came from some air-making machine was very evident, for it issued in puffs from circular openings in the walls. The light was diffused and was clearly an extension of that fluor system which was already engaging the attention of our European engineers when the filament and lamp were dispensed with. It shone from long cylinders of clear glass which were suspended along the cornices of the passages. So much I had observed when our descent was checked and we were ushered into a large sitting-room, thickly carpeted and well furnished with gilded chairs and sloping sofas which brought back vague memories of Egyptian tombs. The crowd had been dismissed and only the bearded man with two attendants remained. 'Manda' he repeated several times, tapping himself upon the chest. Then he pointed to each of us in turn and repeated the words Maracot, Headley and Scanlan until he had them perfect. He then motioned us to be seated and said a word to one of the attendants, who left the room and returned presently, escorting a very ancient gentleman, white-haired and long-bearded, with a curious conical cap of black cloth upon his head. I should have said that all these folk were dressed in coloured tunics, which extended to their knees, with high boots of fish skin or shagreen. The venerable newcomer was clearly a physician, for he examined each of us in turn, placing his hand upon our brows and closing his own eyes as if receiving a mental impression as to our condition. Apparently he was by no means satisfied, for he shook his head and said a few grave words to Manda. The latter at once sent the attendant out once more, and he brought in a tray of eatables and a flask of wine, which were laid before us. We were too weary to ask ourselves what they were, but we felt the better for the meal. We were then led to another room, where three beds had been prepared, and on one of these I flung myself down. I have a dim recollection of Bill Scanlan coming across and sitting beside me.

'Say, Bo, that jolt of brandy saved my life,' said he. 'But where are we, anyhow?'

'I know no more than you do.'

'Well, I am ready to hit the hay,' he said, sleepily, as he turned to his bed. 'Say, that wine was fine. Thank God, Volstead never got down here.' They were the last words I heard as I sank into the most profound sleep that I can ever recall.

The Maracot Deep by Arthur Conan Doyle 219

Credits & Acknowledgements

Most image copyrights are credited on the relevant pages; additional acknowledgments are listed here:

p1: See page 93; **pp2–3:** Sea monster illustration from *Islandia*, Abraham Ortelius, 1585. © Courtesy of Yale University Collection, 2039559; **p3:** Illustration from page 20 of *L'histoire naturelle des estranges poissons marins*, by Pierre Belon, 1551. © Courtesy of Flickr/peacay; **pp4–5:** A selection of fish and sea creatures in a medieval bestiary, c. 13C. © Courtesy of the British Library/Bridgeman Images; **p13 (background):** See page 17; **p19:** See page 23; **pp26–27:** See page 23; **p30:** See page 29; **p37:** See page 35; **p52:** See page 50; **p53:** See page 50; **p58:** See page 61; **p64:** See page 63; **p71:** 'With my cross-bow I shot the Albatross.' Illustration by Gustave Doré, 1875. © Courtesy of iStock; **p81:** An illustration of a mermaid with a mirror and comb from *Les Fais et les Dis des Romains et de autres gens*, c. 1470. © Courtesy of GetArchive; **p82:** See page 93; **p83:** An illustration of two siren-like dolphins from *Hortus Sanitatis*, by Jacobus Meydenbach, 1491. © Courtesy of the Wellcome Collection; **p84:** An illustration of a mermaid from *Smithfield Decretals*, Raymond of Peñafort, c. 1300–1340. © Wikipedia Commons; **p88:** Incidental illustration from the Contents page of *The Poems of Edgar Allan Poe*, illustrated by William Heath Robinson, 1900. © Courtesy of Internet Archive; **p98 (background):** See page 97; **p113:** See page 118; **p129:** 'Only one of its arms wriggled in the air, brandishing the victim like a feather', *from Twenty Thousand Leagues Under the Sea*. Illustrated by Alphonse de Neuville and Édouard Riou, 1871. © Courtesy of Wikimedia Commons; **pp138–39:** Giant squid or 'devil-fish' beached at Catalina, Trinity Bay in Newfoundland, reported in *Harper's Bazaar*, 24 September 1877. © Courtesy of Wikimedia Commons; **p174:** See page 168; **pp178–79:** See page 168; **p182:** See page 181; **p183:** See page 184; **p193:** Illustration from page 40 of *Historiæ animalium liber II*, by Conradi Gesneri, 1587. © Courtesy of Wikimedia Commons; **pp202–03:** See page 9; **p217:** Illustration from page 347 of *Historiæ animalium liber II*, by Conradi Gesneri, 1587. © Courtesy of Wikimedia Commons; **p218:** Illustration from page 119 of *Historiæ animalium liber II*, by Conradi Gesneri, 1587. © Courtesy of Wikimedia Commons

Index

A
Acushnet 91
Aeneas 28–31
The Aeneid (Vergil) 29–31
Ahab, Captain 91
Aldington, Richard 163, 181
Alecton 8, 123
American Review 87
Apsû 6, 11, 12–17
Arkham House 187
Aronnax, Professor Pierre 123–31
'At Ithaca' (H.D.) 181, 185
Athena 29
Atlantic Monthly 97
Augustus, Caesar 29, 33

B
Babylon, *Enūma Eliš* 6, 10–17
Barton, Otis 133, 205
Bathysphere 133, 205
Beebe, William 133, 205
Behemoth 97
Beowulf (Unknown) 51, 52–5
the Bible 91
    The Book of Jonah 47, 48–9, 97
Bloch, Robert 187
The Book of Jonah (Unknown) 47, 48–9, 97
Broadway Journal 87
Browning, Elizabeth Barrett 101
Browning, Robert 101–7
    'Caliban upon Setebos' 101, 102–7
    Dramatis Personae 101
Butler, Samuel 19–27
    Odyssey 19, 20–7

C
'Caliban upon Setebos' (Browning) 101, 102–7
Challenger Deep 205
Challenger Expedition (1872–76) 205
Charybdis 19–27
Christ 6, 47
Circe 19–27, 33–7
'The City in the Sea' (Poe) 87, 88–9
Coleridge, Samuel Taylor 67–79
    Lyrical Ballads 67
    The Rime of the Ancient Mariner 9, 67, 68–79
Confederation Poets 147
'cosmic horror' 187
Cthulhu 187, 189

D
'Dagon' (Lovecraft) 187, 188–92
Death 67
Derleth, August 187
Dido, Queen of Carthage 29–31
dolphins 39
Doolittle, Hilda *see* H.D.
Doré, Gustave 46, 66, 67, 69, 72, 73
Doyle, Arthur Conan 8, 205–19
    The Maracot Deep 9, 205, 206–19
    Sherlock Holmes 205

E
Emerson, Ralph Waldo 97
Enūma Eliš (Unknown) 6, 11, 12–17
Everybody's Magazine 147
Exeter Book 163
Exhibition Universelle (1867) 123

F
fish 39
'The Fourth Map of Asia' 16–17

G
Geats 51–5
Gesneri, Conradi 61, 193, 217, 218
giant squid 8
    Twenty Thousand Leagues Under the Sea (Verne) 123, 124–31
Gilliatt 109–21
Glaucus 33–7
God 47
Greek mythology 63
Grendel 51–5

H
Haeckel, Ernst 113, 118
Haploteuthis ferox 133
Hassam, Childe 97
Hawthorne, Nathaniel 97
H.D. 163, 181–5
    'At Ithaca' 181, 185
    Coterie 181
    Des Imagistes: An Anthology 181
    Helidora and Other Poems 181
    'Sea-Heroes' 181, 182
    'Thetis' 181, 183–4
Hebrew Bible, The Book of Jonah (Unknown) 46–9, 97
Hodgson, William Hope 8, 167–79
    'The Thing in the Weeds' 167, 168–79
Homer 19–27
    Iliad 19, 181
    Odyssey 19, 20–7, 33, 181

Houghton Mifflin Co 29–31
Howard, Robert E 187
Hrothgar 51–5
Hugo, Victor 8, 109–121
   Les Travailleurs de la Mer 109
   Toilers of the Sea 109, 110–21

I
Iliad (Homer) 19, 181
Imagist movement 163, 181
'In the Abyss' (Wells) 133, 134–45
Ishmael 91

J
Jonah 6, 47, 97
The Book of Jonah 47, 48–9

K
King James Bible 47–9
The Knickerbocker 91
kraken 8
   'The Kraken' (Tennyson) 81, 82
   Moby-Dick (Melville) 91
   The Natural History of Norway (Pontoppidan) 57–61, 81, 91
   'The Sea Raiders' (Wells) 133
Kulullû 11

L
Lambert, WG 11–17
   The Babylonian Epic of Creation 11, 12–17
Land, Ned 123–31
Lang, Andrew 51–5
   The Red Book of Animal Stories 51, 52–5
Laocoön 29–31
Leviathan
   Hebrew Bible 97
   'Leviathan' (Thaxter) 97, 98–9
Lovecraft, HP 8, 187–203

'The Call of Cthulhu' 81, 187
'Dagon' 187, 188–92
The Outsider and Others 187
The Shadow Over Innsmouth 187
'Supernatural Horror in Literature' 167
'The Temple' 187, 193–203

M
Magasin d'éducation et de recreation 123
Maracot, Professor 205
The Maracot Deep (Doyle) 205, 206–19
Marduk 11, 14
Maury, MF 123
Melville, Herman 8, 91–5
   Moby-Dick 8, 9, 57, 91, 92–5, 97
Mercier, Lewis 123
mermaids 6, 39, 57, 63
   'The Mermaid' (Tennyson) 81, 84
   'The Merman' (Tennyson) 81, 83
Metamorphoses (Ovid) 33, 34–7
Minerva 29–31
Miranda 101
Moby-Dick (Melville) 8, 9, 57, 91, 92–5, 97
Mocha Dick 91
Montoni, Signor 63
Mussolini, Benito 163

N
Napoleon III 109
The Natural History of Norway (Pontoppidan) 8, 57, 58–61, 91
Naturalis Historia (Pliny the Elder) 6, 39, 40–5
Nautilus 123–31
Nemo, Captain 123–31
Neptune 63
nereids (mermaids) 39, 63

The New Age 163
Nineveh 11, 47
Nordic mythology 8
Nowell Codex 51

O
Octavian 33
octopus, *Toilers of the Sea* (Hugo) 109, 110–21
Odysseus, King of Ithaca 19–27
Odyssey (Homer) 19, 20–7, 33, 181
Ortelius, Abraham 2–3, 41, 42, 43, 166
Ovid (Publius Ovidius Naso) 33–7
   Metamorphoses 6, 8, 33, 34–7

P
Paley, William 101
Pearson's Magazine 133
Phaeacians 19–27
The Physical Geography of the Sea 123
Pliny the Elder 39, 40–5
   Naturalis Historia 6, 39, 40–5
Plongeur 123
Poe, Edgar Allan 8, 87–9
   'The City in the Sea' 87, 88–9
   Tamerlane and Minor Poems 87
   Tamerlane and Other Poems 87
Poetry: A Magazine of Verse 181
Polyphemus 19
Pompei 39
Pontoppidan, Erik 8, 57, 58–61, 81, 91
   The Natural History of Norway 8, 57, 58–61, 91
Poseidon 19
Pound, Ezra 163–5, 181
   Pavannes and Divagations 181
   'The Seafarer' 163, 164–5
Prospero 101
Ptolemy, King 6

Q
Quran 47

R
Radcliffe, Ann 63–5
    The Mysteries of Udolpho 63
    The Sea-Nymph 63, 64–5
Radcliffe, William 63
Rectina 39
Riley, Henry T, *The Metamorphoses of Ovid* 33–7
The Rime of the Ancient Mariner (Coleridge) 9, 67, 68–79
Roberts, Sir Charles GD 147–61
    Earth's Enigmas 147
    Eyes of the Wilderness 147
    The Haunters of the Silence: A Book on Animal Life 147, 149
    Orion, and Other Poems 147
    'The Terror of the Sea Caves' 147, 148–61
Romanticism 63, 67, 147
Romulus 29

S
St. Aubert, Emily 63
Sargasso Sea 167
Sargassum algae 167
The Saturday Evening Post 205
Scherie 19–27
Scylla 19–27, 33–7
'Sea-Heroes' (H.D.) 181, 182
'The Sea-Nymph' (Radcliffe) 63, 64–5
'The Sea Raiders' (Wells) 133
sea serpents, *The Natural History of Norway* (Pontoppidan) 57–61
'The Seafarer' (Pound) 163, 164–5
Setebos 101–7
Shakespeare, William 91
    The Tempest 101
Southern Literary Messenger 87

squid, giant 8
    Twenty Thousand Leagues Under the Sea (Verne) 123, 124–31
Steenstrup, Japetus 8
The Strand Magazine 205
Sumero-Akkadian Cuneiform 11
Sycorax 101

T
Tanach 46–9
Tarshish 47
'The Temple' (Lovecraft) 187, 193–203
Tennyson, Alfred 81–5
    'The Kraken' 81, 82
    'The Mermaid' 81, 84
    'The Merman' 81, 83
    Poems, Chiefly Lyrical 81
'The Terror of the Sea Caves' (Roberts) 147, 148–61
Thaxter, Celia 97–9
    Drift-weed 97
    'Leviathan' 97, 98–9
'Thetis' (H.D.) 181, 183–4
'The Thing in the Weeds' (Hodgson) 167, 168–79
Tia-mat 6, 11, 12–17
Toilers of the Sea (Hugo) 108–21
tritons 39
Trojan horse 29
Trojan War 19
Troy 19
Twenty Thousand Leagues Under the Sea (Verne) 8, 57, 123, 124–31

V
The Vagrant 187
Vergetio, Angelo 9, 93
Vergil (Publius Vergilius Maro/Virgil) 29–31
    The Aeneid 29–31

    Eclogues 29
    Georgics 29
Verne, Jules 8, 123–31
    Twenty Thousand Leagues Under the Sea 8, 57, 123, 124–31
Vesuvius, Mount 39
Victorian era 63, 81, 101, 123

W
W. H. Hodgson's School of Physical Culture 167
Walter, FP 123
Wandrei, Donald 187
The Weekly Sun Literary Supplement 133
'weird fiction' 187
Weird Tales 187
Wells, HG 8, 133–45
    'In the Abyss' 133, 134–45
    The Plattner Story and Others 133
    'The Sea Raiders' (Wells) 133
    whales 8, 39
    Jonah and the Whale 6, 46–9, 97
    Moby-Dick (Melville) 8, 9, 57, 91, 92–5, 97
Williams, Theodore C, *Aeneid* 29–31
Wordsworth, William 67, 81
    Lyrical Ballads 67

Y
Yom Kippur 47
Yūnus ibn Mattā 47

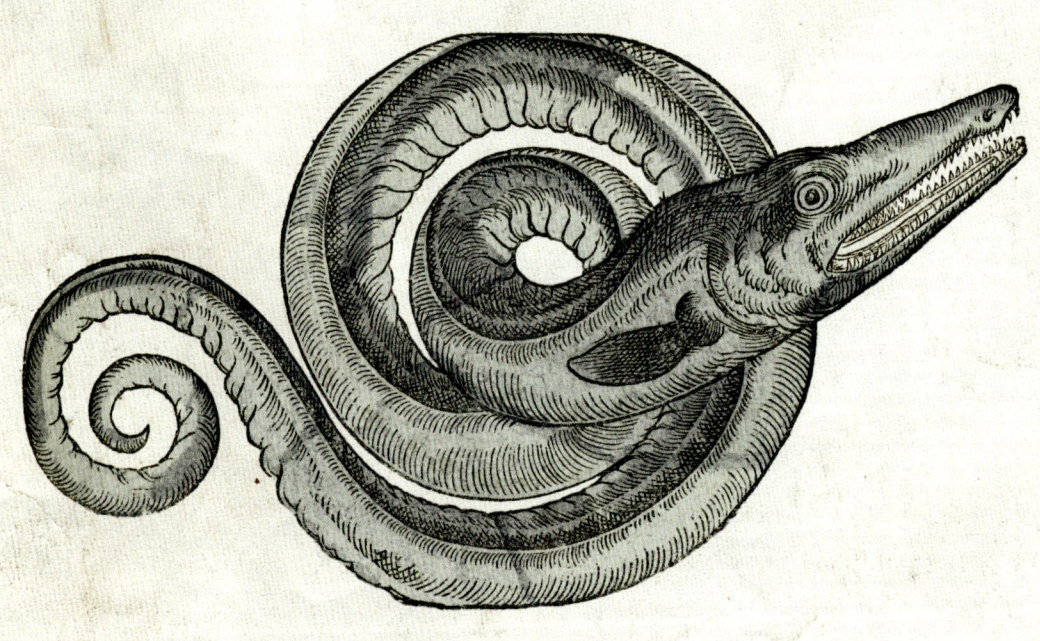